The Endangered Earthlings' Handbook

Opal,
Thank you for everything

The Endangered Earthlings' Handbook

By Team Endangered Earthlings:
Pamela Dawn Erickson
Paul Hollis
Steven P. Sutherland

Endangered Earthlings, Inc.

501(c)3 nonprofit dedicated to reducing our carbon footprint

Copyright © 2019 by Endangered Earthlings, Inc.
All Rights Reserved

No part of this publication may be reproduced, stored in a retrieval system, or transmitted, in any form or by any means, electronic, mechanical, photocopying, recording, or otherwise, without the written permission of the authors.

ISBN: 978-1695-4620-90

Cover Design by Lauren Bowman

Printed in the United States of America.

For our children

Table of Contents

Chapter 1: The Endangered Earthling 1

Chapter 2: Earth 19

    Geosphere 20

    Fossil Fuels 23

    Offshore Oil Drilling 23

    Fracking 32

    The Bad News 35

    Mining 46

    The Cost of Inaction 74

Chapter 3: Hydrosphere/Hydrologic Cycle 76

    Water 80

    Ocean 99

    Overfishing 99

    Climate Change 102

    More Dead Zones 104

    Pollution 107

    Thermal Pollution 108

    Acoustic Pollution 111

    Acidification 113

    Coral Reefs 114

    Cryosphere 115

Chapter 4: Biosphere 120

    Your Home 121

    Your Car 122

Your Groceries 123
Trees 126
   Personal and Social Benefits.. 129
   Community Benefits 130

Chapter 5: Atmosphere 139
Chapter 6: The Virus Human 157
Human Population 157
Food 158
Plastic 164

Chapter 7: Species Extinction 172
Anthropocene Annihilation 172
Conclusions: 191

Chapter 8: Climate Change... 195
Chapter 9: My Earth Ministry....................................... 215
Chapter 10: Solutions.. 235
Climate Change Solutions: 236
Deforestation 241
Hemp 242
   Hemp Part: Seeds 242
   Hemp Seed Hearts 242
   Hemp Part: Stalk 243
   Hemp Part: Leaves 244
   Hemp Part: Roots 244
Bamboo 244
Recycled Paper 245

Wood Composites from Used Plastic	246
Recycled Plastic	246
Cork	247
Soy Plywood	248
Nutshells	249
Abrasive Cleaning	249
Water Filtration	249
Lost Circulation	249
Straw	250
Energy	250
Solar	252
Hydrogen Energy	253
Wind Power	257
Wave Power	259
Hemp Batteries	260
Food	260
Hydroponics	261
Aquaponics	262
LED Hydroponics	262
Urban Eco Villages	263
Cellular Agriculture (#CleanMeatBaby!)	263
Medicine	266
Sound Therapy	269
Ocean Pollution	269
Acoustic Pollution	269

- Thermal Pollution 272
- Ghost Nets 274
- Plastic 274
  - Plastic recycling 275
  - Circular Economy 275
  - Eco-packaging 275
- Sewage and Septic 278
- Species Extinction 279
- Waste Management 284
  - Design 285
  - Recirculate 285
  - Regenerate 286
- Water 287
  - DIY Cleaning Solutions 288
  - Natural Body Care 289
  - Natural Shampoo 289
  - One-Stop Shopping 290
  - Electrolyzed Water 290
  - Off Grid Box 290
  - Water from The Sun 291
- Air 291
- Indigenous Living 292
- Our Village 297

Epilogue 304

Chapter 1: The Endangered Earthling

At this moment, I'm standing on the edge of a dirt cliff created when a developer cleared 100 acres of Florida rainforest to build human habitats. I'm looking down at the tiny finger of trees allowed to stand because they protect Little Black Creek natural springs. As I climb down the dirt mountain and over plastic fencing, there to keep the boundaries between humans and nature, the first being I encounter is a young maple tree. It has been broken in half by a bull dozer, not cut, just pushed half-way out of existence.

Pamela Dawn

As though a single cell of a virus became conscious of the life it was extinguishing, I became aware of my own human condition. I find myself loathing my own species, as a disease annihilating thousands of other innocent species for no other reason than ego. I believe it is possible to live within nature without destroying it, but that doesn't create conditions easily controlled by luxuries or municipalities. For this reason, over-grown, natural yards are not allowed. Where do I belong? Am I no longer a natural being or is there some place where I can find my natural essence? Am I destined to live in opposition with the entire natural world because I am human? Perhaps I must forgive my own human-ness in order to help our species heal and awaken to our tremendous potential to love and to help life to survive *us*.

If there was one thing I could say to you to hold you here with me while I show you my mind, what would it be? Shall I entertain you? Shall I attempt to arouse your curiosity with tears and pounding fists? If we were face to face, I could use my secret weapon usually reserved for that moment in a conversation when the other person is so mad trying to prove their point that you actually see flames rising from the top of their heads. If the individual is all hopped up on ego and hate, I gradually...slowly...ever-so-gently slip my finger right up a nostril, wipe it on their chest and say, "It's okay, I got it." The reactions range from utter fury to humiliation to hilarity, always ending with a snicker...my original intent.

It's my Columbo finger. "What a remarkable tie...just one more thing, sir." Yes, it's disgusting, but what wouldn't I do in the name of comedy? Very little. I'll have you know, the index finger of my right hand has been around, son. My victims: Sidney Ponson (then lead pitcher for the Orioles), an Italian mafia boss, and once, a Middle-Eastern royal whose bodyguard had to look away for laughing. Let's face it; comedy is one of the few things humans got right here. Where were we? Ah yes, me getting your attention.

My loves, me and my greasy finger are just gonna sit here, hold your hand (after I wash mine) and ask about your day. In turn, I want you to stay with me while I stand on the table, shout, "Look at this shit!" and dump out the contents of my brain. Of course, when I am finished, I will sit quietly and listen to your response with a freshly opened mind. (Hunt me down on social media and let's talk.)

I'll dump it on these pages and hope to meet you on some common ground that hasn't been polluted, washed away by rising tides or covered over with plastic.

But this is not *just* my story, nor could I tell you what I want you to know, alone. We, Endangered Earthlings Inc. are a team. Although this began as my personal ministry, it has grown into a collective of brilliant and like-minds. I am going to share a great deal about what brought me to you, to become *your* Pamela. But I am only a part of this collective voice. Please meet our team:

Pamela Dawn, Founder – Environmental Evangelist, Pamela considers herself a philosopher and a warrior for Gaia. But her defining characteristics are her magnetic personality and her unselfish ability to bring out the best in people. There are over seven billion of us wandering the Earth searching for purpose. She has given herself the mandate to bring us together, person to person, people to people, as one indomitable force to save the ground we inhabit as human beings. She's a superhero because that's what superheroes do; they save the world.

Paul Hollis, Executive Director – Retired IBM, Paul is an American author of fictional terrorism and espionage. His #1 bestselling books in The Hollow Man Series follow a U.S. government analyst in a trilogy of suspense across Europe. He is humble, brilliant and a helluva guy.

Steven Sutherland, M.S., R.G., P.G., C.E.M., EE Science Officer – Hazardous Materials Manager/Geologist, Steve has multiple degrees and over 20 years of experience in the environmental and water resources fields, and currently works as Hazardous Materials Manager and Geologist. As Geologists go, Steve's a BAMF!

John Hedgecoth, Creative Director – TV/Film Actor, John is an American actor known for his roles in "Knight of Quixote" and "Murder Comes to Town". Although John is known primarily as a character actor of dark, villainous types, he has shown his versatility in such roles as Bobby Bennet, a fast-talking TV pitchman, a Shaman in the NatGeo series "Origins", and as God in the indie short film "Endgame".

Cisco Coleman, Entertainment Manager – ex NYPD officer, currently Cisco opens doors and windows for anyone with a voice and a message so they can be heard. Cisco Coleman (AKA) DJ Cisco is a 35-year DJ, Music Producer, remixer and Engineer and a Music Promoter. He has a degree in Recording Arts/ Engineering from the Center for The Media Arts. He owns The DJ Cisco Radio Network LLC. Under the DJ Cisco Radio Network LLC, is his record label: DJC Radio Records Global and DJC Radio Global an internet radio network/show and founder of the. Project Unity Tour: a movement to feed the homeless and give artists a platform to perform, to be seen and heard and to win industry prizes.

ENDANGERED EARTHLINGS, INC.

Paul Valdes, Extreme Weather Crisis Response Consultant - Firefighter (retired) Paul Valdes Jr, a native New Yorker, was a first responder to the events of 9-11 and spent several days assisting the rescue and recovery missions. Former Army 82nd airborne, America's Guard of Honor 25th ID Hawaii Military Police, he is a true superhero. His 90 days of service in the Gulf Coast following Hurricane Katrina is only one of the many, many ways he has bravely and honorably served us all; too many to list here but you can learn more about him on our website. As our mission is proactive, we help in every way we can by assisting and serving survivors of extreme weather events. Paul consulted a team of Endangered Earthlings who served in the Florida panhandle following hurricane Michael and continues his service as our Emergency Response Team leader. He is a badass mofo, but one of the kindest and gentlest humans you will ever meet. We are honored and grateful to have him with us on the Endangered Earthlings Crisis Response Team.

As individual members of a team or unit, we don't always agree. I believe this gives us a unique position from which to reach average human beings. Just as each individual opinion and contribution is what makes up this lovely world of people, each of us Endangered Earthlings have a perspective that is valid, relevant and necessary. Admittedly, I am a reluctant humanist. But if I fail to believe in our ability to find a common ground on which to stand as a species, then all is lost. Humans are that single thing which must change in order to save our planet. Our message to you is to speak, think and move as a community and so this is our collective minds in just that form.

Although you will hear the most personal and intimate details from me, about my life and why I have devoted it to our planet; the experts on our team will fill in some of the most relevant facts about the science of our dying planet, as well as what we agree are our best solutions. You will have me at a disadvantage, as I will show you my most sacred self: the memories which created me. Please be kind and gentle with them. Nothing has ever wounded me so badly as a betrayal. I want you to trust me, so I am taking this huge risk; making what is a tremendous effort for me to instill trust inside these words. I intend to carve off pieces of self and hand them to you. Why? Because I believe we just don't have the time to fuddle with egos, labels or politics. Yes. There are things I won't talk about. If the telling of my story injures someone else, then it is not my story to tell. And I'm not going to tell the whole feckin' blah blah blah, just the parts that are important to my relationship with you.

Nor will I tell you about how; when I was 16 my father began the descent into death of Agent Orange poisoning. Ugh. I guess that I must tell you as it brought me to you. (Grabs another glass of wine and box of tissue)

My parents had divorced and Mom remarried his best friend when I was 12, but my father's impending demise was devastating. I flew to Ohio where he had been working for NASA on the first Space Shuttle launch and said my goodbyes at his bedside. As I held the hand of this tiny stranger; not the mountain of a father who raised me into my preteen years, I remembered all the things that make us cry with sentimentality in our old age. I reminisced on him taking me to work with him now and again. He was a Tech Sargent in the Air Force who served as a radar tech in Vietnam's DMZ. After he came home, on the weekends he worked as a mechanic for the Thunderbirds. We had magical weekends camping out in hangars, having exciting adventures in flying machines. I will never forget when my sister CayDee and I got to fly in an open cockpit plane. He did gut dropping negative g's and loopty loops. I'm sure my mother was having a coronary but I was hopelessly addicted. Every time I got the chance, I would go to work with my dad on the trainer; a mobile simulator unit on a train that moved from base to base for training. The last time he took me, I was nine and I got to fly through a storm in the B52 simulator. It was the coolest thing I'd ever seen.

As I held the hand of my dying father and listened to him groan in pain, I thought of the many times as a toddler that I would awaken in the night and find him stretched out in the dark in his easy chair. I would climb up in his lap, burrow my face and tiny fingers into the mats of hair on his chest and listen to the same stories along with the click, click, click of the slides from Vietnam. "This little boy's name was Ben, (Binh) and this was the school we built for his village. This is what it looked like after they blew it up. See all those white things in the field? That's toilet paper. We built them toilets but people got blown up in them so they were afraid to use them. See that building? That was our barracks. Your Daddy's a coward you know. I got up in the middle of the night and headed to the bunker when it got too loud. Took a piece of shrapnel right up the center of my bunk. I lost all my good friends that night. Shoulda been me." I still smell the mix of beer and roasted peanuts on his breath and the sweat that was now pouring off his head running down dark brown curls and I would give anything to smell them again. Those terrible, wonderful things would prepare me for all the Vietnam vets who would become my Dads later.

I left that broken image of my dying Daddy and I returned to my beloved California. I met an older man on the beach and three months later, married him. I had a great stepfather but my Mom kept him at an arm's length from us. I say us, but my sister was married the same year my Mom remarried, so it was just me. I'm sure she had her reasons. But I guess I was desperate for someone to watch over me, like the song says. Unbeknownst to me, I had married into one of the most feared crime families in the country. A week after we were married my father in law handed me a very weathered paperback book and said, "This is your bible now." Honestly and regrettably, all I remember now is the word Mafia written in red. Read it I did. He was right. It changed what I knew about the world. It was the rich cultural beginnings of an army of outlaws, forced into crime to protect homes and family. But what changed me the most was my life with his son. His son was nothing like the wise and gentle man who created him. Even when I heard the father speaking in hushed tones in the living room to his partner about things I imagined only happened in movies; I found nothing but love, adoration and respect for him. Okay, FBI guys, no I never heard anything that will help you solve a crime. Everything I overheard began with, "So did you hear that…?" and I can't recall a single story that makes any sense today.

His son, however, was something from a horror film. Now this is where I will pause. I can feel my chest tighten, my stomach turn and I am a powerless child, held in torturous captivity again. Maybe I can tell you more tomorrow…

Okay, it's the next day and I've got my sea legs back. Here we go: two years into a life of abuse no human should ever endure, let alone daily for 28 months, I found the courage that only becoming a mother inspires and I ran for our lives. Luckily, the man I married fell out of favor with the family, so we were allowed to drift in the breeze and begin life again. She saved my life…my little baby girl. But I was so very, very damaged and broken. Over many years, previously untreated injuries were identified, mostly by the improper healing of bones. Each time I heard, "We noticed a previous fracture…" I remembered and relived each of the horrifying injuries. It's remarkable the way our super cool brains have a way of tucking things neatly in corners of our unconscious until we are ready to digest them. An operation to fix my permanently stuffy nose should have taken two hours, but the severity of the previous injuries caused it to take six. There were seven breaks I could recall by the pop and momentary blindness. Then, there were two busted ribs, a fractured pelvis and a place in my brain indicative of multiple concussions. He loved to hit me on the back of the head because it didn't leave marks. There were almost daily sexual assaults, especially through my pregnancy. I was isolated from friends and family but the worst of it all was the demoralizing, "No one will ever love you…want you…care about you…you're fat and ugly." Those voices still pop up in my daily life. You'd think half a century of life would have me further along in this healing process but as the lovely song says, "You Can't Rush Your Healing." Maybe the open wounds keep me just strong enough to keep my fists up when I need them. Okay, enough of that. I can't move forward looking back.

The only place my little savior and I could go when we were free was home. My Papa, the designation for my stepdad, had retired as a Master Sergeant from the Air Force and was working his way up the ranks with

Honeywell in Reston, Virginia as a Project Manager on the FA-18 flight simulator. I spent my 18th year tucking in my wee one and heading to the plant to fly the Hornet in a $3 million simulator. It was all I wanted to do. I called a Marine recruiter who met me at the pet store where I worked part-time and we started the paperwork. We got to a designation, I said, "I'm gonna fly the Hornet." He laughed until he realized I was dead serious. My guts fell out when he told me women couldn't fly combat.

Around that same time, my parents took us to the 'wall' for the first time: The Vietnam Memorial Wall in Washington, D.C. Walking the Wall changed the course of my life.

As I descended the ominous black granite cliff, I couldn't help but reach out to the first names I saw. My fingers could feel them as my brain processed the letters that were once beloved sons and fathers. Immediately, I realize two of the names were related; father and son. My heart was already screaming for that lonely woman left with only half a soul to wander the Earth. That black mirror began as a tiny triangle, fit for just a finger then it grew like a shadow, foretelling all the death still to come in front of me. It was physically oppressive; as if the spirits of soldiers pressed out of the shadow to touch the life they were denied. By the time my head was fully submerged in the black reality of war, I was sobbing. I was swirling in a dark whirlpool, realizing that the profound experience of having my father ripped away from my life for some political maneuvering was not just my untimely loss; it was suffered by innumerable fellow Americans. I hurt for them. I hurt for the young boys who never fell in love or conquered middle age. I hurt for the ones who lead valiant battles, knowing full-well there would be no victory no matter the outcome. I suffered to the core of my soul for all of us. I passed by grown men with grey beards, crumpled onto one knee, touching a single name and wailing in silence. I passed by teddy bears, challenge coins and letters written by tiny hands. There was no light or hope or oxygen and I felt as though I was dying with them. I wanted to look away and walk swiftly to the end of my own suffering but how could I? Each name demanded that I look at it with respect, reverence and gratitude. I looked, but with shame. Somehow, my species did this to them all. I looked. I sobbed. I walked reverently as though my feet were passing through deep chasms of Dante's world and finally, an end to it. I emerged completely drained; exhausted by the battle and defeated. I didn't look into eyes. I didn't greet the passers-by with my self-imposed duty to smile at the world. I skulked up the walkway like a gangplank to certain doom. I heard a deep voice from far away…the table I had grabbed to hold my soul inside my trembling body. "Did you know they are still there?" The man who spoke was Michael Hagen, a returned vet who sold his business, bought a motorhome and devoted his life to bringing home his forgotten brothers: 2,646 to be exact. I will fast forward here, as this is not a biography. Michael became my father. He taught me that the only thing which kept him fighting was the man next to him. This made all the soldiers his brothers, including my dad. So, to him, in a very real sense, I was his child too. That was the day my healing began.

I became involved in the POW/MIA issue. I marched side by side with Michael and all the dads I would come to know. Among them was John (Top) Holland, Bobby Garwood, Eugene Simmons, Adrian Cronauer, (Good Morning Vietnam) and many more that would shape my life for the coming decade. They made me strong. They taught me how to live with the ghosts of my PTSD and to use them to become my own sort of warrior. I would go on to become the POW/MIA representative to Gov. Norman Bangerter of Utah under the guidance of an amazing man named Dale Madsen and would eventually become the first female in the state of Utah to receive the Amvets Certificate of Merit.

Eventually, Michael handed over the title of National Chairman of the Campaign for Freedom. I had become a full-fledged activist.

I was a struggling divorced mother with a baby, attending Merced College full time, and I devoted every remaining moment to educating the public and gathering the strays (lost veterans) to bring their brothers, my fathers' home. I was nineteen and too determined not to be afraid of anything. Besides, my fathers had made quite a lethal little soldier out of me; so I took terrible risks and met the strays in the dark corners all by myself more than once. I was lucky to have escaped my youthful exuberance unscathed. I remember one such occasion. A fellow student who stopped by my information booth mentioned an uncle who, like my father and so many others, had never really come home at all. The PTSD kept him bound to the hell of the disembodied memories, so he remained isolated in his elderly mother's home, where she too lived the hellish nightmare. She frequently woke to her son's fingers wrapped around her throat as he screamed vulgarities into her terrified eyes. I entered the home alone and he reluctantly escorted me to the 1950's mint green dinette set and pulled out a chair and slid it across the floor to me. "You think you know what the FUCK I've been through? You and your little white bread cheesy smile." Before I could answer he flew into a rage about his first beer in country. He stepped off the chopper and the old-timers, the twenty-year olds who had survived the first seven months of tour, said, "Let's get you that first in-country beer." They tugged him to a tower of dead bodies, lopped off the decaying head of a V.C. soldier, scooped out the soupy brains and dumped the can of beer in; "Welcome to country, boy!"

When he finished raging, I sat tall and said, "Do you know what it's like to be dragged by your hair 30 feet through gravel then raped and sodomized?" He was too shocked by my calm composure to answer and just glared back. I said, "We all have our hells to contend with. Let's talk about how you can help us." He became one of the most active members of our little truth campaign, a changed man and one more father.

All my Dads, the ones who shaped me and the ones I have not yet met, give me the courage which now sustains me. I have Forest Lync Brusoe as one of my enduring dads to thank for reminding me I'm not an orphan to this day. I love you, Pop.

My days as an activist did not end when I left the POW movement. Many years later, while running a green construction consulting firm that focused on sustainability principles, I was interviewed for a local talk-show in a studio built in the back of a book store in Jacksonville, Florida. The host introduced me to the store's owner, Dorothy Pitman-Hughes. When I saw the photo and realized who she was, it was like being at the Lincoln Memorial in Washington D.C. and having Lincoln rise to his feet and shake my hand. Okay, awkward but flattering to meet this amazing woman. Dorothy is a piece of history that I watched in adoration from the sidelines throughout the solidification of the feminist movement.

My mother was forced to evolve before her time when Daddy went to war. I can only describe her as Jackie Kennedy-ish; the quintessential Donna Reed housewife, always perfectly quaffed who dressed my sister and I in bonnets and gloves and was an etiquette ninja. When my father went off to war, she was forced to enter the work-force. I have come by my determined disposition quite honestly, as she is the kind who never does anything half-way. My sister and I watched our mother transform from demure lady of the house, to brazen and bold business woman. This was around the time that Dorothy and Gloria Steinem were forging a new possible life for us females, and my mother taught us their principals like mantras. I was only 6 or 7, but she sat us in front of the news at night and showed us their faces and their courage, and made us understand that we can be anything we can imagine because they had fought the worst of the battles for us. War was a concept with which I had regrettably become familiar, since I prayed every night for my Daddy to survive his.

It was one of my life's most auspicious honors to meet Dorothy in the flesh, and each time I have spent time with her since, I have come away changed and more awake.

At the time Dorothy and I met, my business was going so well that I was able to hire my Vet-Dad Adrian Cronauer to be our spokesperson. Then the shit hit the fan...all the shit on the entire fan…direct hit. I was a construction consultant in 2008 when the great collapse began.

Almost immediately, the dominoes of my life began to fall. The daily stress meant that hours were spent in front of the computer, feverishly trying to save our business which triggered muscle spasms and, in turn, exacerbated arthritis that kept me up at night. The pain was unbearable. In an attempt to avoid narcotics, I put a pill in my mouth called Cymbalta. A few days later, I had what seemed to be a grand mal seizure, except that I was fully conscious the whole time. Having remarried four years prior, it was my husband who stood helpless over me. For hours, I watched the horror in his eyes as doctors, one by one, came and went. "Sir, there is nothing wrong with your wife. This is not a seizure."

After two years of misdiagnosis, I would finally receive the right diagnosis of Generalized Dystonia, but as an incurable condition, it offered my life no hope of change. The condition had rendered me helpless by these increasing body spasms until I ended, finally, in a hospital bed in my home. I didn't mean, ended up. I mean ended. I lost my business, had to send my little girl to live with her father, and one Sunday my husband went to the grocery store and never came home. My neighbor was a home nurse and she was a strong but loving black woman who I still call "Mamma." Lillian Washington remains dearer to me than if she was my own blood. She came in the next day and said, "Honey, Jack ain't comin' back. Said he couldn't watch it no more."

I was a Cali girl stranded in southern Georgia, with the nearest family member two hours away in Florida, and much too proud and stubborn to make anyone know how bad my condition had become; my secret hell. My darkest night had begun. At its worst, this condition locked me in for hours and sometimes days. I looked like a stroke victim and my limbs postured and jerked like palsy. When it was really dark, sometimes I was locked in completely catatonic, but horrifyingly awake and fully lucid on the inside. Mamma Lillian only came to check on me a few times a week, so the cruelest fights were wrestled in complete isolation. Sometimes, I would wake up with a terrible headache on the floor in the hallway or the bathroom. I remember a particularly bad episode in which I was left to contend with my own weakening mind, slowly losing a will to live. My eyes were fixed on the wall in front of me. My head and jaw were twisted, my right arm repeatedly banging into the rail. I watched the shadows move from light to dark to light again, pathetically uttering, "Home, pweez god, home... pweez gawwwd...home...," for hours and hours until the desperate prayer came out in a squeaky whisper. I could feel the tears streaming, but I was not even allowed the merciful release of a sad expression...just frozen and begging for death.

Merciful death did not come and the slow battle to allow myself to live, drudged on. Some months later, Mamma Lillian took me to have injections into my back, where a fatal car accident (another story, another time) had caused a vertebral fracture. As was a common occurrence, I went into a full dystonic episode as a reaction to propofol. The doctor attending was new to the practice and I am certain my seeing him was divinely inspired, as this man was a neurologist (Dr. Kai McGreevy, now of Riverside Pain Management.) Before I had regained consciousness from the sedation, the tachycardia started. In the hell of my life alone, this was my signal to get to a safe place and lay down. I usually had about 30 seconds. But this time, I was unconscious and in the hands of a gifted neurologist and he was about to save me. When the dramatic event was over, he asked me what had been prescribed in the past. I told him, Benadryl. He seemed to be outraged that I had never been prescribed a medication for movement disorders. That afternoon, he wrote a prescription for Sinemet, Carbidopa. On the way home, Mamma ran in and filled it and the next morning I took the first dose. Two hours later, the toes on my left foot, which had been curled under for months, released to a normal position. After the next dose, my postured hand relaxed to look like a hand again. I had full control of my body for the first time in what seemed like years. It was over. It took weeks of this new reality to prove itself permanent in my life, but this miracle of miracles finally convinced me that I was not over. As Rainer Maria Rilke puts it, "Life has not forgotten you, it holds you in its hands, it will not let you fall." I was like a child in Candy Land. Life was the unimaginable prospect Dorothy had offered me when I was a little girl. The air was sweet, the sky a perfect blue, and nothing could take my hope from me. Nikos Kazantzakis nailed it when he said, "Hope is our greatest temptation. It is the stuff of our deepest desires and too often, the blade of our most devastating disappointments."

There is a caveat to this happy ending. The change had not come in time to save the wreckage of my life. It was three years since he left and my husband had moved on and was ready for a divorce. I was waiting for the answer to my disability application, but it is such a long process, that it was no life saver either. And the life which had been restored by medication came with the unhappy side effect of vomiting for an hour or two each time I took it, three times a day; so gainful employment was not an option. I had no choice but to put my belongings in storage and move in with a family member, my oldest daughter. I struggled each day to find my next chapter, the next crusade; to fulfill life with some greater purpose. I even filled out an application for a life of service as a nun, despite the fact that I wasn't Catholic. (Giggling) Surely the church had room for one more hopeful agnostic.

Although my disability application was denied, despite the 480 pages of documentation by physicians, I made a plan. With the little bit of money from my divorce settlement, I rented a little trailer that my Aunt Marcy and I could afford and I set a plan in motion. All the terrifying tidbits of information about our dying planet, which I had learned while writing the commercial for Adrian, had never left my mind. Once on my feet, I headed straight for Dorothy. She reminded me who I was and of what I was capable. She gave me direction and lovely, lovely FIRE! I put those things together and here I am.

There is a story I need to tell you about Dorothy that will paint a portrait of her contributions better than my shero worship. Ms. Dorothy had a meeting in south Florida with Hillary Clinton while she was campaigning. Ms. Dorothy is not a fan of flying, so she, I and her daughter Delithia rode a bus together. As we waited for the bus, I made sure Dorothy and Delithia were comfortable in the little café and stepped outside to confirm her arrival time. When I came back in, there was a crowd gathered around her and several people waiting to get nearer to her. One by one, they came from behind the counter, from the baggage

Pamela Dawn with Dorothy Pitman-Hughes

loading dock, from the ticket counter, waiting room and drivers stepped off their busses to come and shake her hand. Each one imparted deeply personal thanks for the great work she had done for African Americans and told her how she affected their lives and their families. I'm utterly sobbing to recall it. And you know, the most beautiful part is her response. She almost looked embarrassed as she shook their hands, asked their names and had something personal to tell them about what she knew about their families or their counties. This is leadership. This is courage and this is how I learned to love.

I was standing 10 feet away when Hillary Clinton said to her, "Dorothy, you have done so much to get us here. Thank you."

Dorothy Pitman-Hughes is one of the biggest reasons I am here and writing this to you today. Please read all you can find on her amazing life and you will come to love her as I do.

Bus Trip to Meeting with Hillary Clinton

My loves, I am here as an amalgamation of the people I love and who have loved me. Deeply ingrained in that mix, are the natural creatures, plants and pets that have sustained my mental health and healed my soul after each confrontation of darkness. I have no choice but to stand here completely exposed and offer myself to you. How could you love me unless I first give you my love? I am yours. Love me. Heed me. Destroy me (or try.) I give you that power. But please, please, please read the words on these pages and take them to heart. This is me loving you.

Of all the things I've ever done, being a mother was my favorite job. It was more important, meaningful, fulfilling and life altering thing I've ever done. Every choice after my first child was made for my child and each one to follow. When I imagine our Mother Earth, the Mother of all life on our planet, I can't imagine what she is thinking right now, how

Hillary Clinton & Dorothy Pitman-Hughes

she is feeling, or even how she can possibly sleep. One of her children, the collective of humans, is killing all the rest. We are destroying her body and her children are pointing multiple weapons at their own heads. Maybe it is that thing my Dads created in me, the warrior combined with the mother which made me rise up out of my own ashes and rush to the front lines of this environmental battle to speak for those whose language is a vibration, a grunt, a chirp or a slow grumbling roar only felt on the bottom of our feet. But I have given the remainder of my days on Earth to speak to you in ways that will awaken you to the remarkable brilliance of your purpose here. Of all the species to ever exist on this magnificent planet, ours is the first with the power to determine how we will evolve. If you believe the educated minds about to speak to you in this book, your empirical reasoning will cause a shift in your logic paradigm and your choices will change. If you believe me when I say that you are not a stranger to me, you are my kin, you are me, just a few genes separated, that you are someone I love: If you believe that as I type these words there are tears rolling down my face and something resembling a prayer bursting out of my solar plexus that you will…then you and I have much work to do as kindred souls. Trust me. Stay with me long enough to read the words to come. If I lose you in the logic, STAY. If I offend you with my candor, STAY. If I repulse you with my oozy gooey hope or severely Irish-American potty mouth when I say that we can fix the fucked-up mess we made of this place; forgive me, and let me love you anyway. The absolute singular hope we have to survive ourselves is to find a common ground on which to build trust. From there, we can do anything. I am telling you my secrets and showing you all my cards because what is at stake is bigger than anything socially acceptable that I could have been. Reaching you will be the second most important thing any human has ever done. The most important? I'll defer you to the movie City Slickers:

Curly: Do you know what the secret of life is?

[holds up one finger]

Curly: This.

Mitch: Your finger?

Curly: One thing. Just one thing. You stick to that and the rest don't mean shit.

Mitch: But, what is the "one thing?"

Curly: [smiles] That's what *you* have to find out.

I'll meet you in the lobby for popcorn.

Chapter 2: Earth

Spoiler alert; one of my evil tactics in an attempt to convince you to love thy planet as thyself, is to anthropomorphize her (i.e. try to paint her with human attributes) so you can relate to her and not want to destroy her. It's the same idea behind the genetic programming which causes a father to recognize himself in his offspring and not want to eat it.

Humans, meet Planet Earth:

Earth

If we compare planet Earth to our own human body, we will see similarities:

- Geosphere, Her skeletal system
- Hydrosphere, Her circulatory system
- Biosphere, Her skin (This is where we live. So, we're like mites but eating way more than our fair share of dead skin.)
- Atmosphere, Her respiratory system.

Geosphere

Earth has a Geosphere that is akin to our skeletal system. It does all kinds of cool things like self-regulate in ways that we see as volcanoes, Earthquakes, fissures, geysers, mountains, islands and the shifting of continents. Imagine if we could self-heal our old bones and regenerate/restructure them. She's been doing it all her life.

We all know the baby of the family has ways of shaking things up and wreaking havoc, not only on mom's body but the whole family. And we baby humans are severely messing up her game.

More than a thousand years ago, humans discovered uses for a nasty smelling, greasy black stuff in the ground. Vacationing in Galveston, Texas, as a kid in the 70's, I will never forget going home from the beach with black feet. Not the country-cute she's been walkin-all-day-barefoot black; I mean thick, sticky, tar-coated BLACK. It was incredibly gross and nearly impossible to get off skin. Forget getting it out of cloth; we just threw the towels and swimsuits away. All that Texas offshore oil drilling sent clumps of the stuff washing up on the beaches. But, at some point in history some clever human set fire to it and if you've ever watched how long a Frito corn-chip will burn, you can imagine their excitement. How could they imagine that empires, societies, and human civilization itself would be built on the back of its trade, or foresee the millions of lives that would be sacrificed to get it?

The earliest known oil wells were drilled in China in 347 AD or thereabouts. They were able to penetrate depths of 800 feet. But the real commercialization of petroleum as an industry began on American soil. In 1875, a man named David Beaty discovered crude oil at his home in Warren, Pennsylvania. Eventually, this became the Bradford oil field and by the 1880s, it produced 77% of the global oil supply. And although it was driven by the demand for kerosene, the industry grew through the 1800s; right into the automotive industry we know today.

In response to rocketing global oil prices driven by the Organization of the Petroleum Exporting Countries (OPEC), on December 22, 1975 congress passed the Energy Policy and Conservation Act of 1975 (EPCA.).[1] By preventing domestic producers from getting crude oil out of the country, the intent was to increase domestic supply and keep domestic prices down. And because America imported much more oil than it could have exported, domestic producers didn't feel they were missing out. December 22, 1975, a United States Act of Congress responded to the 1973 oil crisis by creating a comprehensive approach to federal energy policy. The primary goals of EPCA were to increase energy production and supply, reduce energy demand, provide energy efficiency, and give the executive branch additional powers to respond to disruptions in energy supply.

Around this time, oil took the country of Ecuador from one of the poorest in Latin America to a thriving economy which continues to rely on oil for 40% of its export earnings; most of which comes from the northeastern Amazon basin known as El Oriente. A thriving economy is great, but what has been the cost? Let's take a walk over to Ecuador for a minute so we can see the distinct hazards of the oil business.

In the 1970's and through the 80's, by land policies, the Ecuadorian national government shifted 4.5% of the country's population away from this region. This included eight tribes of indigenous peoples and many impoverished citizens. The El Oriente region covers 100,000 square kilometers (km^2) of a tropical rainforest at the headwaters of the intricate Amazon River network. This region is one of the most biodiverse regions in the world, containing species of plants and animals we may have never seen. Today, it is a matrix of roads, pipelines and oil facility. As for the indigenous locals and migrants; the expendable masses? New York attorney Judith Kimerling, acting as National Resource Defense Council's (NRDC) Latin American representative, enumerated the environmental toll in her 1991 exposé, Amazon Crude. Drilling operations produced 3.2 million gallons of wastewater a day, which the

---

[1] The Week: America's oil export ban, explained
https://theweek.com/articles/595142/americas-oil-export-ban-explained

oil companies stored in unlined pits that leached toxins into the soil and waterways. Equally as damaging was the estimated 16.8 million gallons of oil spilled from pipelines in the region.[2] The contamination of water essential for the daily activities of thousands of people has resulted in an epidemic of cancer, miscarriages, birth defects, and other ailments."[3] In May 2010, The Rainforest Action Network referred to the industrial pollution in Ecuador as "one of the largest environmental disasters in history".

Thankfully, on April 26, 2019, hundreds of Ecuadorian Waorani tribesmen and women took to the streets of Puyo, the regional capital of Pastaza, rejoicing a new court ruling which determined that the Ecuadorian government could no longer auction their land for oil exploration without their consent.

If we measure the damage to the Amazon in terms that directly affect you, we have to take a look at what trees mean in the biosphere and what the disruption of biodiversity means to human existence (those chapters coming up). Oil is not the only destroyer of our precious Amazon. Animal agriculture is still the primary cause of deforestation. But sit tight. You'll hear about those topics shortly. For now, let's take a look at what, in 2019, made the USA the big man on the oil campus.

Remember back in the 70's…or a few paragraphs ago, when we learned that oil was once a crisis in this country? Specifically, in October 1973, the members of Organization of Arab Petroleum Exporting Countries or the OAPEC (consisting of the Arab members of OPEC) enacted an oil embargo "in response to the U.S. decision to re-supply the Israeli military" during the Yom Kippur war, which lasted until March 1974. In the late 60's and into the early 70's, the U.S. oil boom ended as production fell. International oil prices surged. In response to stagnant

---

[2] A Village in Ecuador's Amazon Fights for Life as Oil Wells Move In
https://www.nrdc.org/onearth/village-ecuadors-amazon-fights-life-oil-wells-move
[3] Business and Human Rights Centre: Human rights impacts of oil pollution: Ecuador
https://www.business-humanrights.org/en/human-rights-impacts-of-oil-pollution-ecuador-22

growth and price inflation during this period spawned the term, stagflation. Although many still argue that the crisis was merely the perception of crisis, it was enough to send the powers that be scrambling into solutions mode. By the 80's, prices began to stabilize. But the chaos sent many back to the drawing board, trying to surmise a solution to that pesky, impermeable shale that stood between the oil industry and oceans of Texas tea and soon the next boom would come as the sound of hydraulic fracturing, or fracking. But we continue to see offshore and traditional oil wells pop up.

Fossil Fuels

We have a long history with traditional oil hunting, via land oil rigs and offshore drilling. Yes, there are benefits; such as:

- Jobs
- Enterprise
- Energy security

But what is the cost?

Offshore Oil Drilling

Before we deep dive into the facts, please note this: Yes, there are a great deal of articles referenced that may not be in your reading list, but please know that the information was curated for you based on if, and only if, there were peer reviewed studies sited. Here we go.

Deep in the floor of the ocean lie pockets of oil. Until recent years, offshore drilling was somewhat limited to the continental shelf. This is the seabed around a continent which is shallow enough for a rig to drill. However, technology has made it easier for deep water drilling to occur, as well.

So how is this done? First, they have to find the pockets. According to the National Energy Technology Laboratory, "Deepwater Technology".

Offshore reservoirs can be identified through 3D seismic surveys, which scan the seabed to understand the rock formations. Sound waves are sent, received and decoded from the survey vessels to create 3D images, which depict the oil and natural gas pockets hidden between the porous rocks underneath the sea.[4]

You may have noticed the horrific news of thousands of sea mammals washing up dead. Many were found to be starving with bellies full of plastic, but in April 2012, hundreds of dead dolphins washed ashore with injuries that indicated acoustic impact:

"...evidence of middle- and inner-ear damage, lung lesions and bubbles in the blood are consistent with acoustic impact and decompression syndrome, leading to speculation that oil exploration in the region may be to blame."[5]

"May be to blame," You think? This leads us to the more important issues than the benefits of oil: What does it cost Earth and all life on her?

Offshore rigs can dump tons of drilling fluid, metal cuttings, including toxic metals, such as lead chromium and mercury, as well as carcinogens, such as benzene, into the ocean[6]

To better understand the impacts of offshore drilling, let's take a look at the components and risks. The process of extracting oil from the ocean floor includes things like:

- Drilling muds
- Produced water
- Deck runoff water
- Flowline and pipeline leaks

---

[4] National Energy and Technology Laboratory: OFFSHORE TECHNOLOGY http://www.netl.doe.gov/research/oil-and-gas/deepwater-technologies
[5] Ecology Today: Cause of 3,000 Dolphin Deaths in Peru Likely to Remain a Mystery https://www.ecology.com/2012/04/19/3000-dolphins-dead-peru/
[6] Oceana Protecting the World's Oceans: BP Deepwater Horizon Oil Disaster by the Numbers https://usa.oceana.org/impacts-offshore-drilling

- Well failures
- Catastrophic spills

Each of these components comes with some harmful effects for humans and marine life as well as long-term contamination of the environment. Drilling mud (also called drilling fluid) in petroleum engineering is a heavy, viscous fluid mixture that is used in oil and gas drilling operations to carry rock cuttings to the surface and also to lubricate and cool the drill bit.[7]

Regarding drilling mud, Oceana says:

Drilling muds are used for the lubrication and cooling of the drill bit and pipe. The muds also remove the cuttings that come from the bottom of the oil well and help prevent blowouts by acting as a sealant. There are different types of drilling muds used in oil drilling operations, but all release toxic chemicals that can affect marine life. One drilling platform normally drills between seventy and one hundred wells and discharges more than 90,000 metric tons of drilling fluids and metal cuttings into the ocean.[8]

Produced water is the fluid trapped underground that comes up with oil and gas. Nearly 20% of the waste associated with offshore drilling is produced water and usually has an oil content of 30 to 40 parts per million. For the Cook Inlet in Alaska, that means 2 billion gallons of produced water is released into the inlet each year containing around 70,000 gallons of oil.

In a comprehensive study conducted by 24 scientists for Frontiers in Environmental Science, the experts stated:

Discharges of water-based and low-toxicity oil-based drilling muds and produced water can extend over 2 km, while the ecological impacts at

---

[7] Encyclopedia Britannia: Drilling Mud
https://www.britannica.com/technology/drilling-mud
[8] Oceana Protecting the World's Oceans: : BP Deepwater Horizon Oil Disaster by the Numbers https://usa.oceana.org/impacts-offshore-drilling

the population and community levels on the seafloor are most commonly on the order of 200–300 m from their source. These impacts may persist in the deep sea for many years and likely longer for its more fragile ecosystems, such as cold-water corals.[9]

When oil reaches the natural seas via deck runoff, pipeline leaks, and catastrophic spills or even leakage from routine maintenance of ships, it spreads predominantly on the surface. If the slick gets caught in a storm, it moves like a deadly shadow over great distances, devastating all life in its path.

The description of this moving slick by Water Encyclopedia is more horrifying than I could paint with poetic imagery:

Oil that contains volatile organic compounds partially evaporates, losing between 20 and 40% of its mass and becoming denser and more viscous (i.e., more resistant to flow). A small percentage of oil may dissolve in the water. The oil residue also can disperse almost invisibly in the water or form a thick mousse with the water. Part of the oil waste may sink with suspended particulate matter, and the remainder eventually congeals into sticky tar balls. Over time, oil waste weathers (deteriorates) and disintegrates by means of photolysis (decomposition by sunlight) and biodegradation (decomposition due to microorganisms). The rate of biodegradation depends on the availability of nutrients, oxygen, and microorganisms, as well as temperature.

Oil Spill Interaction with Shoreline.

If oil waste reaches the shoreline or coast, it interacts with sediments such as beach sand and gravel, rocks and boulders, vegetation, and terrestrial habitats of both wildlife and humans, causing erosion as well as contamination. Waves, water currents, and wind move the oil onto shore with the surf and tide.

---

[9] Frontiers in Environmental Science: Environmental Impacts of the Deep-Water Oil and Gas Industry: A Review to Guide Management Strategies
https://www.frontiersin.org/articles/10.3389/fenvs.2016.00058/full

Beach sand and gravel saturated with oil may be unable to protect and nurture normal vegetation and populations of the substrate biomass. Rocks and boulders coated with sticky residue interfere with recreational uses of the shoreline and can be toxic to coastal wildlife.[10]

**A dead, oil-covered bird near the Houston Ship Channel.
(Melissa Phillip / Associated Press)**

For fur-bearing mammals like sea otters, oil destroys its insulating abilities and it makes bird's feathers incapable of repelling water. This exposes them to the bitter elements and they die from hyperthermia. For many birds, as they try to clean themselves, they ingest the oil which poisons them. When oil is mixed into the water column, adult fish suffer enlarged livers, changes in heart and respiration rates, fin erosion and impaired reproduction, as well as increasing mortality of eggs and larva. These are not characters in a book or creatures from another planet. These are our Earthling-siblings. Our lives depend on theirs, whereas all life on Earth would flourish without humans. If there is a single consciousness aware of our existence, I'm certain this thought has

---

[10] Water Encyclopedia: Oil Spills: Impact on the Ocean
http://www.waterencyclopedia.com/Oc-Po/Oil-Spills-Impact-on-the-Ocean.html#ixzz5pX4XXjFE

crossed their awareness, but we should be the ones truly contemplating this truth.

On April 20th, 2010, the sun rose in the Gulf of Mexico on the worst environmental disaster in United States history. By the time the world knew what was happening the oil rig, Deepwater Horizon, was fully engulfed in flames and her crew was only beginning to comprehend the scope of the catastrophe.

Chief Electronics Technician, Mike Williams was one of the last workers to leave the rig as he relentlessly tried to save as many of his fellow crew members he could.

"...there was such chaos. Once we realized that – that there were 11 men unaccounted for – it was impossible to go back. There was no way to get back up there, no way to go back and try to help them. That's probably the hardest thing to deal with."[11]

What went wrong on the Deepwater Horizon? In an article by New Scientist, BP claims that there were eight failures that caused the calamity.

Approximately 70 km from the U.S. gulf coast, crews aboard the Deepwater Horizon were preparing to abandon a well. Cement was dumped in the borehole to seal it and prevent leakage of oil. But the cement did not hold and oil began leaking into the surface pipe. This failure was the first of the ominous eight.

Besides the shoddy cement formula, the second thing to fail were the valves intended to shut off the flow of oil and gas to the surfaces. The valves failed.

Third, while conducting tests to determine if the well was in fact, sealed, the test results were misjudged and the leak went unnoticed.

---

[11] PEOPLE.COM: Inside the Harrowing True Story of Deepwater Horizon Survivor Mike Williams: 'It Haunts Me' https://people.com/movies/deepwater-horizon-the-true-story-of-survivor-mike-williams/

It is left to the judgement of the crew to detect and notice if oil and gas are flowing towards the surface by scrutinizing well pressure. About 50 minutes before the rig exploded, an increase in pressure was noted, but not interpreted as a leak, making this the fourth failure.

About eight minutes before the explosion, mud and gas spilled out onto the floor of the rig. The crew immediately acted to close the blowout preventer, a valve over the borehole; but this valve, too, failed, becoming failure number five.

The sixth failure: Instead of diverting the mud and gas over the side of the rig, the crew opted to run the flow through a device onboard called a separator. The mechanism was designed to separate small amounts of mud and gas but swiftly overloaded causing flammable gas to overwhelm the rig.

Although the rig was equipped with an onboard gas detection system that should have sounded the alarm and triggered the closure of ventilation fans to prevent the gas reaching potential causes of ignition, such as the rig's engines, it failed. Number seven.

The eighth and final blow for Deepwater Horizon was the BOP, or Blow Out Preventer. The explosion destroyed the control lines the crew was using to attempt to close safety valves in the BOP. But, as part of its own safety mechanism, two separate systems should have shut the valves automatically when it lost contact with the surface. One system had a flat battery and the other a defective switch. Consequently, the blowout preventer did not close.[12]

Chief Electronics Technician, Mike Williams spent nearly two years shut-in, suffering from severe PTSD. But despite the many triggers involved with the retelling of the story in the movie, *Deepwater Horizon*, based on this horrific piece of our American history, Williams showed up

---

[12] New Scientist: The eight failures that caused the Gulf oil spill
https://www.newscientist.com/article/dn19425-the-eight-failures-that-caused-the-gulf-oil-spill/#ixzz5uRGXmee0

on set every day for his fellow crew members stating,

"It's an important tribute to my 11 brothers. When I agreed to assist them with this project, it was under the direction of 'I have to speak for 11 people who cannot speak,' " he says. *"I have to tell these guys how to get this right so that their image is held up in the highest light possible. That was my motivation for the entire project."*

My loves, that word "deregulation" has personal and deadly consequences for our species, not just the natural world. Anyone you know who has a job is protected by the regulations put in place by the catastrophes that preceded them. After deep suffering, the voices of their loved ones rose up and were heard by the lawmakers who built walls around human life to keep it safe and thriving. Yes, the deregulation of industry makes it more profitable, providing more jobs to make more things we may or may not need, but those jobs, when the safety measures are removed or deregulated, are each a life at risk. No one should expect to go to work and not come home, regardless of the hazards.

Regarding the Deepwater Horizon incident and consequent lawsuit:

On September 4, 2014, U.S. District Judge Carl Barbier ruled BP was guilty of gross negligence and willful misconduct under the Clean Water Act (CWA). He described BP's actions as "reckless," while he said Transocean's and Halliburton's actions were "negligent." He apportioned 67% of the blame for the spill to BP, 30% to Transocean, and 3% to Halliburton. Fines would be apportioned commensurate with the degree of negligence of the parties, measured against the number of barrels of oil spilled. Under the Clean Water Act fines can be based on a cost per barrel of up to $4,300, at the discretion of the judge. The number of barrels was in dispute at the conclusion of the trial with BP arguing 2.5 million barrels were spilled over the 87 days the spill lasted, while the court found that 4.2 million barrels were spilled. BP issued a statement disagreeing with the finding, and saying the court's decision would be appealed.

POWER2CHOOSE™ ALTERNATIVE ENERGY

"The Path To a Brighter Future will be Lit By A Lamp that Burns Nothing..."

Rocky Buldo is a man on a mission to change how we think about, acquire and use energy. He was born in a time and raised in the place where Edison built the foundations of the electronics industry. His personal goal is to complete Edison's work on SAFE DC power for digital age devices and pass along an alternative source of power to humanity.

The current power grid is the indisputable master of AC power and that's not going to change. We can only choose to change what we do. But how do we change what we do if there is no choice in the AC equation? We must either take it or leave it and suffer the consequences. Where there is no alternative option, there can be no free market choice without the Power2Choose™ Alternative Energy.

Rocky Buldo has created the Power2Choose™ Alternative Energy that can provide at least enough off-grid power to deal with our own footprint in the internet of things. This system of power and load is backward compatible to any power grid in the world, but dependent on none, and forward compatible with any device that has a plug. Our mission is to build a bottom up free market solution because it's not going to come from the top down.

Power2Choose™ Alternative Energy typically collects energy from the sun but can be used with any power source such as wind or water. Energy is stored in DC batteries, then distributed to household appliances, directly to those digital devices that can innately accept DC power sources and through a inverter for older AC only equipment still in the house.

You are invited to join in a revolution of self-empowerment from the stone age to the digital age: (https://enoughtechnologies.org).

In July 2015 BP reached an $18.7bn settlement with the U.S. government, the states of Alabama, Florida, Louisiana, Mississippi and Texas, as well as 400 local authorities. BP's costs for the clean-up, environmental and economic damages and penalties had reached $54bn. In January 2018 a detailed estimate of the "Ultimate Costs of the Oil Spill", published in the Journal of Corporate Accounting and Finance, amounted to 145.93 billion USD.[13]

I wish the Deepwater Horizon was the end of our oil woes. But, sadly, with the most recent deregulation of the industry, fossil fuels are still going strong. Fracking is just beginning to dip into decline after years of boom status. But we are still in the midst of fossil fuel distress.

I bet you didn't know that right now, there is an active oil spill in the gulf that's spewing as much as 4,500 gallons a day and it's been going on for 15 years. In 2004, Hurricane Ivan caused an off-shore oil platform in the Gulf of Mexico collapse into the sea and there appears to be no end in sight. It would seem that to cap the remaining 16 wells that are leaking, would cause more harm than good. So, the U.S. Coast Guard installed a separate containment system, which has been collecting some 30,000 gallons of oil over 30 days but this fix is a bandage on a surgical wound. Who will foot the bill for the long-term fix? Taylor Energy Company shut down in 2008 so does that put the burden of correction in the pockets of us taxpayers?

Oil is not our future; it is our destructive past…time to move on. "But wait!" You say. "Fracking made us energy independent!" Did it? Did it really?

Fracking

What is fracking? Picture this. Your best friend needs your life-saving bone marrow and you're a donor match. Just as they put the sleepy-gas

---

[13] Wikipedia: Deepwater Horizon explosion
https://en.wikipedia.org/wiki/Deepwater_Horizon_explosion#Casualties

mask on your face, the doctor says,

"I'm plunging a needle deep into your bone where the marrow is, injecting a bunch of nasty chemicals with great force and blowing the marrow out of tubes at the other end of your bone while simultaneously fracturing your bones from the inside."

Fracking.

The end.

Okay, not really but you get the picture. Eventually, the structure of your skeleton would completely fail (i.e. Fracking and sinkholes). Not to mention, every system in your body would be poisoned by the chemicals. And that, my friends, is fracking.

Shale formations underlie much of our United States and are the source of natural gas and oil. Shale is a dense sedimentary rock. Although shale is very porous, it is not very permeable, meaning; it is difficult for liquid or gas to pass through it. This allows the shale to hold a lot of oil/natural gas but makes it difficult to remove these substances. Fracking makes it possible to extract the oil and natural gas trapped in deep within shale formations.

Fracking is the process of injecting fluid, (often dangerous chemicals) into an underground rock formation using high pressure to open fissures, or cracks, and allow trapped gas or crude oil to flow through a pipe to a wellhead at the surface. This practice is used often in the United States and although it has made a vast amount of natural gas available to energy companies, the cost in terms of public health and environment has been disturbing, to say the least.

Nonetheless, thanks to fracking, team America has taken the Iron Throne in the kingdom of fossil fuels. Having devised an effective method for making unreachable resources available, the old supply and demand game has done its thing and consumers have won. The demand for fossil fuels and natural gas goes up with a flourishing economy and the increase of population. Next chapter: "How Barry White Effects Our Economy," subtitle, *Bow Chica Waow Waow*. But that increase is somewhat gradual. The supply, however, has increased with a boom as a result of fracking. By increasing the supply of crude oil and natural gas, the cost to distributors has been driven down which passes the savings on to we consumers. Fracking saves Americans $180 billion annually on gasoline, according to Forbes Magazine as the U.S. economy paid $390 billion for crude oil and finished product imports in 2008; while, in 2016 the U.S. paid out $78 billion for oil and finished products -- a decline of $312 billion.[14].

Fracking has produced 1.7 million jobs. Although the study by the U.S. Chamber of Commerce entitled, *U.S. Chamber's fracking job boom: Behind the numbers* produces impressive numbers in terms of jobs, it also highlights the fact that,

Skeptics with environmental and citizens groups have questioned the numbers and also the benefits that these jobs actually provide to local communities. Many industry jobs are not filled by local residents, and a boom town effect, including escalating cost of living and other social problems, has been documented in places where an extraction industry rapidly arises.[15]

There have been jobs created, such as:

- Construction
- Oil and gas extraction
- Metal fabrication
- Truck transport.

---

[14] Forbes: Fracking Saves Americans $180 Billion
https://www.forbes.com/sites/rrapier/2017/06/02/fracking-saves-consumers-180-billion-annually-on-gasoline/#2f49aa3612dd

[15] GlobalEnergyInsitute.org: U.S. Chamber's Fracking Job Boom Behind Numbers
https://www.globalenergyinstitute.org/us-chamber%E2%80%99s-fracking-job-boom-behind-numbers

Besides the jobs created, the stimulation of local economies that comes with new industry, the 21 participating states benefit when big oil brings big bucks in terms of taxation. Way down on the private sector, the drilling companies owe private land owners leasing fees as well as royalties on gross production at no less than 12.5%. But, when you're the big man on a little campus, you make the rules. As Don Feusner, a Pennsylvania rancher found out the hard way. Somewhere, buried in pages of legalese, Chesapeake Energy, the company that drilled his wells, was retained nearly 90% of Feusner's share of the income. The explanation was as clear as sunset in a whiteout snow storm: Unspecified "gathering" expenses. To have those inexplicable fees detailed would require a lengthy lawsuit and the legal fees only an oil giant could afford to pay. Considering the fact that 30% of drilling takes place on federal land, I bet those checks are never short. But as Propublica.org points out, in states like Pennsylvania:

Pennsylvania attorneys say many of their clients' leases do not allow landowners to audit gas companies to verify their accounting. Even landowners allowed to conduct such audits could have to shell out tens of thousands of dollars to do so.[16]

For certain states and municipalities, fracking is good business. Not so much for the private landowners.

As of this publication, the U.S. is the #1 producer of crude oil. We are extracting fossil fuels from the bones of Mother Earth at the rate of 12 million barrels per day. At the rate of consumption around 20m BPD, we are still not energy independent nor are we consuming our own crude. Here is where we segue to the downfalls of fracking.

The Bad News

A recurring theme in this book is our diminishing and finite resources.

---

[16] ProPublica: Unfair Share: How Oil and Gas Drillers Avoid Paying Royalties
https://www.propublica.org/article/unfair-share-how-oil-and-gas-drillers-avoid-paying-royalties

By an increase in population and flagrant overuse, we are using more water fresh water than this planet can provide. Right now, there are resource wars taking place and this fact will come to your door if we're not careful. On that note, current third generation fracking wells use between 10 and 30 million gallons of water *per* well.

The water used in fracking is treated with chemicals, as many as 1,000 different chemicals are used, some radioactive; and they do things like:

- Thicken the water
- Prevent mineral buildup in the pipe
- Reduce friction.

Fracking water is treated with chemicals you may recognize such as arsenic, benzene, cadmium, lead, formaldehyde; but there are many more lethal chemicals you may not. Let's just look at one: Benzene.

Benzene, or benzol, is a highly flammable liquid which evaporates rapidly into air and dissolves slightly in water. The sweet odor of benzene presents at approximately 60 parts of benzene per million parts of air (ppm) and the taste in water at 0.5 - 4.5 ppm. (1 ppm = one drop to forty gallons.) It is found in air, water, and soil and while it from natural sources, compared to industrial, the natural occurrence is negligible unless you're in the path of a volcano or forest fire; sadly, more common these days due to extreme weather events caused by climate change. However, levels of benzene detected in areas dotted by fracking sites are comparable to 10 volcano eruptions at one time on the low end (Rich study) and 100 volcano eruptions on the high end (Macy study). [17]

That wastewater can be a toxic blend that's very difficult to treat, in part because it may contain high levels of corrosive salts, naturally occurring radioactive materials, and fracking chemicals whose identities are

---

[17]From the Styx: Measure benzene, damn it!
https://fromthestyx.wordpress.com/2018/02/12/measure-benzene-damn-it/

considered trade secrets and which even the U.S. Environmental Protection Agency can't list.[18,19]

Improperly installed or poorly maintained, improperly bored or sealed fracking wells have contaminated drinking wells with this toxic water in several sites. But it's okay, the companies were fined. (Sardonic smirk) The problem with extracting precise and usable data as to the extent of contamination is that the private land owners sign non-disclosure agreements; so their wells are untouchable. The non-private land wells? That means the federal government owns that data. As long as it took them to admit that my father's death was a direct result of his Agent Orange poisoning, I wouldn't hold your breath.

In the course of the transportation, storage and recycling of fracking wastewater, the containment of this toxic fluid depends on the vessels and machines which process it. Besides the mechanical failures, this fluid does escape through fractures and failures of wells, tanks and pipes. You can't just filter that shit; it's gone.

Remember that number, 10 to 30 million gallons of water per well? A study by Duke University tells us that,

Much of the water fracking uses is essentially lost to humanity. Either the water doesn't escape the shale formation or, when it does come back to the surface- it is highly saline, is difficult to treat, and is often disposed through deep injection wells.[20]

According to *Scientific American*, **Fracking Can Contaminate Drinking Water,** Former EPA scientist Dominic DiGiulio states:

---

[18] Resilience: Fracking Wastewater Spikes 1,440% in Half Decade, Adding to Dry Regions' Water Woes https://www.resilience.org/stories/2018-08-21/fracking-wastewater-spikes-1440-in-half-decade-adding-to-dry-regions-water-woes/

[19] DESMOG: Fracking Can Contaminate Drinking Water, Has Made Some Water Supplies "Unusable," Long-Awaited EPA Study Concludes https://www.desmogblog.com/2016/12/13/fracking-can-contaminate-drinking-water-has-made-some-water-supplies-unusable-epa-announces-long-awaited-study

[20] Science Advances: The intensification of the water footprint of hydraulic fracturing https://advances.sciencemag.org/content/4/8/eaar5982

We showed that groundwater contamination occurred as a result of hydraulic fracturing," DiGiulio said in an interview. "It contaminated the Wind River formation.[21]

In 2010, Congress commissioned the U.S. Environmental Protection Agency (EPA) to study the impact of fracking on drinking water and finally, June 2015, the U.S. EPA released its long-awaited final draft of its report determining:

The future availability of drinking water sources that are considered fresh in the U.S. will be affected by changes in climate and water use. Declines in surface water resources have already led to increased withdrawals and cumulative net depletions of groundwater in some areas…The colocation of hydraulic fracturing activities with drinking water resources increases the potential for these activities to affect the quality and quantity of current and future drinking water resources.[22]

In 2008, the people of Pavillion, Wyo. noticed a bad taste and smell in their drinking water. U.S. EPA launched an inquiry, led by DiGiulio. Preliminary testing suggested that the groundwater contained toxic chemicals. In 2013, the agency handed over the investigation to state regulators without publishing a final report.

DiGiulio published a peer-reviewed study in Environmental Science and Technology that suggests people's water wells in Pavillion were contaminated with fracking wastes, usually stored in unlined pits dug into the ground.

---

[21] Scientific American: Fracking Can Contaminate Drinking Water https://www.scientificamerican.com/article/fracking-can-contaminate-drinking-water/?redirect=1
[22] EPA's Study of Hydraulic Fracturing and Its Potential Impact on Drinking Water Resources https://www.epa.gov/hfstudy/executive-summary-hydraulic-fracturing-study-draft-assessment-2015

The study also suggests that the entire groundwater resource in the Wind River Basin is contaminated with chemicals linked to hydraulic fracturing, or fracking.[23]

But Pavillion was not the only community affected by fracking contaminants. An article by ThinkProgress.org entitled *Fracking is Destroying America's Water Supply* states:

The game-changing study from Duke University[24] *found that "from 2011 to 2016, the water use per well increased up to 770%." In addition, the toxic wastewater produced in the first year of production jumped up to 1440%.*[25]

As a side note, 70% of fresh water is used by agriculture with an estimated 15% increase expected by 2050. There are better ways to feed 8 billon and we'll talk about that in *Solutions*. Back to fricking fracking.

Let's take a jaunt over to North Dakota. Ah, ND. My stepdad was stationed there for the last 18 months of his USAF career and I learned so much. I was 13 and got my first big girl kiss from the boy next door. I learned that mosquitoes are the ND state bird, the creeks have leaches and you can't go outside when it's 90 below chill factor or your ears, eyeballs and lungs will freeze. Oh for neat, eh? Having moved there from California, needless to say, I was so ready for our 3-month excursion around the coast of Florida in a motorhome.

But on a more serious note, **Environmental Science and Technology** published an article in April 2016 entitled, *Brine Spills Associated with Unconventional Oil Development in North Dakota* by, Nancy E. Lauer, Jennifer S. Harkness, and Avner Vengosh, Division of Earth and Ocean

---

[23] Scientific American: Fracking Can Contaminate Drinking Water
https://www.scientificamerican.com/article/fracking-can-contaminate-drinking-water/?redirect=1
[24] Science Advances: The intensification of the water footprint of hydraulic fracturing
https://advances.sciencemag.org/content/4/8/eaar5982
[25] Think Progress: Fracking is destroying U.S. water supply, warns shocking new study
https://thinkprogress.org/fracking-is-destroying-americas-water-supply-new-study-9cb163923d24/

Sciences, Nicholas School of the Environment, Duke University, Durham, NC, US:

Researchers found high levels of ammonium, selenium, lead and other toxic contaminants as well as high salts in the brine-laden wastewater, which primarily comes from hydraulically fractured oil wells in the Bakken region of western North Dakota[26]

Of then, 9,700 drilled wells there were more than 3,900 **brine spills**. Groundwater, streams and soil were all contaminated. If you are inclined to decipher the specific breakdowns, the tables are very meticulous and can be found in the article referenced above.

In 2015, NASA satellite data showed that 21 of the world's 37 aquafers have passed the sustainability tipping point[27]. The world is literally running out of water. You'll see many references to water throughout this book but with consideration of increasing populations and climate change, we can't afford to lose the billions of gallons wasted through fracking, especially considering the fact that we have viable and affordable alternatives now. (You'll see those in *Solutions.*)

As a matter of fact, the city of Barnhart, Texas ran out of water due to fracking. But, wasted water and the trail of **contamination** it leaves behind is not the only issue to consider here with regard to water. As it turns out, many of the fracking induced Earthquakes are caused by the injection of the wastewater. Not surprisingly, the waste "water" which also includes sand, brine and chemicals, being pumped through the ground at extremely high pressure to crack open rocks can reach faults at risk of breaking and cause it to slip; causing **Earthquakes.**[28] On June 10, 2019, just such an event occurred at Eastlake, OH. Yes…Ohio. And it

---

[26] Science Daily: Contamination in North Dakota linked to fracking spills
https://www.sciencedaily.com/releases/2016/04/160427150617.htm
[27] AGU100: Quantifying renewable groundwater stress with GRACE
https://agupubs.onlinelibrary.wiley.com/doi/full/10.1002/2015WR017349
[28] EOS – Earth and Space Science News: More Earthquakes May Be the Result of Fracking Than We Thought https://eos.org/research-spotlights/more-earthquakes-may-be-the-result-of-fracking-than-we-thought

was a 4.0! I'm from California so that wouldn't even have gotten me out of bed but when your sleeping crust rumbles, listen. Yes, it can and will get worse if the number of wells continue to increase.

Texas and oil go together like coffee and bovine lactation. But the French have a term, Follie a deux. It's a shared psychosis and when the two meet, well you get Bonnie and Clyde: The real, heartless, ruthless, hero-murdering Bonnie and Clyde, not the fabulous version with Faye Dunaway and Warren Beatty. But the repercussions don't stop there. The only way I want to feel the Earth move under my feet when the deck collapses from raucous party peeps. But, fracking and drilling are causing the Earth's crust to just give in. Geophysicists from Southern Methodist University said west Texas has been "punctured like a pin cushion with oil wells and injection wells since the 1940s," causing the ground to rise and fall and the result is **sinkholes**.[29]

Roy Orbison really would be *Crying* if he knew his hometown, Wink, Texas was at risk of being swallowed by greed. Check out this Wink hole:

Sinkhole in Wink, Texas

---

[29] SMU University: Research shows Permian Basin sinkholes are growing https://blog.smu.edu/research/2019/05/22/cbs-7-research-shows-permian-basin-sinkholes-are-growing/

Like oil drilling, I wish I could say the damaging effects of fracking stop there, but alas…there's more bad news about fracking. Let's focus for a moment on the point of this whole process. Aren't we here to extract crude oil and natural gas? Well, guess what, fracking does not produce pure methane. About half is heavy hydrocarbons and that mixture is utterly unusable. What to do with all that nasty gas? Build a natural gas pipeline from every well and hope the plant is not at capacity? The choice most oilers make is Flaring and Venting. Flaring is the burning of natural gas emitted from wells, that can't be processed or sold, while venting is the release of methane gas which occurs at the point of well maintenance, completion as well as well maintenance, pipeline maintenance and tank maintenance.

One of the main chemical byproducts from fracking is methane, a colorless, odorless and flammable gas; 34 times more heat-trapping than CO_2 and it is estimated that at least 4% escapes into the atmosphere in the process of fracking. But the contribution to greenhouse gasses is only part of its virulence. Methane has detrimental impacts on air quality surrounding the fracking site.

Flaring releases an abundance of air pollutants. April 4, 2019, Earthworks.org released an article entitled *Flaring and Venting* that states:

The Ventura County Air Pollution Control District[30], in California has estimated that the following air pollutants may be released from natural gas flares: benzene, formaldehyde, polycyclic aromatic hydrocarbons (PAHs, including naphthalene), acetaldehyde, acrolein, propylene, toluene, xylenes, ethyl benzene and hexane. Researchers in Canada have measured more than 60 air pollutants downwind of natural gas flares.

Whether by wastewater contamination or air pollutants, there are major repercussions for those who live in proximity of fracking sites.

---

[30] VENTURA COUNTY AIR POLLUTION CONTROL DISTRICT: AB 2588 COMBUSTION EMISSION FACTORS http://www.aqmd.gov/docs/default-source/permitting/toxics-emission-factors-from-combustion-process-.pdf?sfvrsn=0

How does fracking affect the health of humans within proximity? April 15, 2019 EHN.org tells us that after a decade of research, here's what scientists know about the health impacts of fracking. (This article sites Oxford Research Encyclopedia of Global Public Health) Fracking has been linked to: preterm births

- High risk pregnancies (low birth weight and or infant with low infant health index)
- Asthma
- Migraine headaches
- Fatigue
- Nasal and sinus symptoms
- Skin disorders.

Other health impacts like cancer and neurogenerative diseases require more time to develop but SITE Pavillion Wyoming, Penn and Texas suits

In Pennsylvania, which accounts for 19% of U.S. natural gas, in consideration of demographic factors such as race and income, the study found that in the county, the number of new wells and the well density corresponded to an increase hospitalization rates for kidney infections, kidney stones, and urinary tract infections, particularly in women ages 20-64.[31]

But darlings, all is not lost. In my quest to understand this complicated but (for now) necessary business, I found a brilliant young account manager for Pioneer Energy, Matt Foster, who presents a viable solution to this gas conundrum. You'll get all their juicy details in the *Solutions* chapter, but in my fifth-grade understanding of the problem, I asked Matt the following question:

---

[31] Environmental Health News: Fracking linked to increased hospitalizations for skin, genital and urinary issues in Pennsylvania https://www.ehn.org/fracking-linked-to-increased-hospitalizations-for-skin-genital-and-urinary-issues-in-pennsylvania-2630649093.html

Pamela: *Can the gas extracted by fracking be used as fuel? Why, why not and what next when it comes out of the ground?*

Matt: *Yes, the gas can be used on the site as fuel most commonly to power generators and compressors although there are other uses.*

The natural gas that is extracted from wells as a result of fracking can be used as fuel, but before it can be, it must first be processed. The gas in raw form is a combination of different constituents. It has some heavier hydrocarbons in it such as ethane, propane, butane, and even heavier components. All these hydrocarbons other than methane are referred to as "natural gas liquids or NGLs." These liquids in the gas cause problems if used in engines of field equipment.

Typically, the raw gas out of the ground is sent through a pipeline to a centralized processing plant which separates it into individual "purity" products which are then sold as individual fuels. Unfortunately, it is not always possible to connect pipeline to the well pads. Sometimes the pipeline is delayed. Sometimes the plant that the pipeline is connected to is already at capacity. In these cases, Pioneer Energy's Flarecatcher systems can instead process the gas (very similar to what would be done at the larger centralized plant) to separate out these NGLs. The result is that the NGLs are now in liquid form. As such, they can be much more easily trucked off site for further processing. What's left after the NGLs are removed is lean, conditioned gas (mostly methane) which is perfect for use as fuel in natural gas generators or compressors.

One interesting thing is that if there is enough raw gas to be processed that we could potentially run big enough generators to not only power the equipment on site but potentially nearby communities.

My loves, if you live near a fracking site, you need Pioneer Energy! I would love to see us move with great speed and efficacy into the age of renewables but it's going to take time. Therefore, the dark side of this industry needs to be something we work on solving and not ignoring. The oil industry will be with us a bit longer, sadly.

ALL of this to make us energy independent, but the sad truth is, we are *not*. All the blustering recently about how clean our air is and how our plastic is not the problem. Yeah, that's because we ship it to the countries getting the worst raps for the refining of *our* oil and the recycling of ou*r* plastic. Check this out:

The United States imports billions of barrels of crude a year, about a third of it from OPEC. At the same time, the United States exports a substantial quantity of the stuff. That's because most of the refineries in the United States were built when the country was still obliged to rely very heavily on imported oil, and so most of them are optimized to handle the "heavy sour" stuff from abroad rather than the "light sweet" stuff from Texas. It is not the case that a barrel of oil is a barrel of oil is a barrel of oil. "Every single molecule from here on out has to be exported," Cynthia Walker of Houston-based Occidental Petroleum told the Texas Tribune.[32]

All the damage to our environment so that the fat cats at the top of the oil industry can get fatter. In the meantime, we little people pay for it with our health and the future viability of our land and fresh water. It's important to understand that there is not just a single kind of oil, there are many and they all create different products. Despite the fact that we're spewing the stuff out, much of that oil is not the right type needed to make products Americans use the most. That makes it impractical to keep American oil in America.

Guys, we have been programmed to live top down by industry for far too long. We have allowed the tail to wag the dog, so to speak. We are more than seven billion average humans being driven hard into consumerism by the very few who sit on the top of the money piles and they have dictated to us what we need and how we get it. But there is a new day upon us. We common folks are realizing that we don't need half the shit they sell us. It stacks up around, beneath us and in our garages until we

---

[32] National Review: Yes, the U.S. Exports Oil, but We're Not Energy Independent
https://www.nationalreview.com/2019/07/yes-the-u-s-exports-oil-but-were-not-energy-independent/

need a storage unit for the excess of *things* we don't use but once every year or two. So, wake up and understand that you and I have had the power all along. When we stop buying the crap they're selling us; when we stop making the crap they sell right back to us so that we can just barely cover our basic essentials; when we begin to live our lives on our terms…well they're gonna fight us. And *then* we will force them to become part of a sustainable human experience. There truly is power in numbers. Use your power! Walk, ride a bike, cyber-commute to a job that works around your life. Buy, (anything but plastic) and live local. Make every change you can make in your daily life until the industry follows your lead.

All of these concepts will be clear and attainable when we get to *Solutions*.

To continue our hard look at our effect on the Geosphere, we must realize that oil and fracking injuries are not the only thing we humans are doing to destroy the skeleton of our beloved planet. There are still horrific mining practices taking place all over the world, including the U.S. and Canada. Our executive director, Paul Hollis, came from a family of coal miners and knows this industry better than me. There is enough you need to know about mining to give it its own chapter so continuing our look at the Geosphere, next read *Mining*, by Paul Hollis. Besides, he's a brilliant bestselling author so I'll let you enjoy a break from my gooberificness.

Mining

By Paul Hollis

Coal is a fossil fuel. It is the altered remains of prehistoric vegetation that originally accumulated in swamps and peat bogs. The energy we get from coal today comes from the energy that plants absorbed from the sun millions of years ago. All living plants store solar energy through a process known as photosynthesis. When plants die, this energy is usually released as the plants decay. Under conditions favorable to coal formation, the decaying process is interrupted, preventing the release of the stored solar energy. The energy is locked into the coal.[33]

Humans mine coal to generate 40% of the world's electricity. Simply put, coal-fired electricity generation is a five-step process:

1. Thermal coal (either black or brown) is pulverized into a fine powder, then burned
2. The resulting heat is used to turn water into steam
3. The steam at very high pressure is then used to spin a turbine, connected to an electrical generator
4. The spinning turbine causes large magnets to turn within copper wire coils; this is the generator
5. The moving magnets cause electrons in the wires to move from one place to another, creating electrical current and producing electricity.[34]

The mining of coal is one of the toughest jobs a person can have, even by today's much improved standards. I come from a long line of miners across the southern regions of the United States. Both grandfathers worked the mine. One died from black lung and the other just quit living somewhere along the way. Many of my uncles followed their fathers in search of black gold, including my own father. That is, until it rained hell one day.

It was the kind of morning kids remember later in life. The sky was a brilliant, crystal blue of sparkling sapphire; not like the washed-out bubble that hangs over the planet today. Birds were awakening and nocturnal species were kissing their babies goodnight. A breeze from the west carried the smell of wild flowers and fir, stirring up the black dust and rust across the stained courtyard that spilled out from the mine's lone entrance.

---

[33] World Coal Association: What is Coal? https://www.worldcoal.org/coal/what-coal
[34] Origin: How does coal make electricity? https://www.originenergy.com.au/blog/what-is-coal/

None of the miners that morning saw any of this and some never would again. They left company-owned homes like vampires under the glow of fading moonlight to bow their heads at the blessed alter of the low elevator that dropped them down to the pit where they earned their pay. Back then, salaries mostly came in the form of cash with automatic deductions for taxes, tools, housing, and the company store which left little to nothing beyond existence. There was always hell to pay after the money was gone, every single day, and miners were on their own to pay that debt with what was left of their bodies. But today would demand an ultimate price.

Switch on the headlamps. Light the lanterns. Lead us with blind faith to the devil's door.

Bare, flickering bulbs led the way twenty feet at a time like bell jars filled with lightening bugs through a maze of worked out tunnels. The sound of boots echoed along the corridors of dark, greasy, almost geometric rock. No one ever counted the steps. No one wanted to know how far they were below the ground.

Dear Lord, just get us back to our families before the dead of this night.

The work rooms opened at right angles off the main heading. Most men were looking forward to digging their ten to twenty feet of coal and getting back home safely. But today's room was already a very dangerous and over-mined thirty feet wide. It was more than five feet beyond prudence; beyond sanity. They had taken as much as this cavern would give. Before they abandoned her though, as a final act they would pillar rob any remaining coal; a technique to extract the last bits of energy by pulling out the supports, allowing the roof to collapse onto the mine floor.

By the time the four miners picked up their sledgehammers, the ceiling was already raining cinders of coal. It was coming down soon, maybe on its own. They had to remove a key pillar or two and get out of there before the whole thing came down. My father swung at the first piling like Lou Gehrig in the bottom of the ninth, connecting just below the shim and joist. No one could tell for sure if the loud crack came before or after the hammer hit. The 8X8 crossbeam dropped suddenly and hit him across the face. That's when the roof fell on the unforgiven.

Bad news travels faster than a bullet to the heart, and leaves the same wound.

The rumble shook the Earth above. That wasn't unusual. After all, the shake of underground black powder explosions was common but when my mother looked up from her chores, I knew something was different this time. She ran. She ran through the screen door and down the dirt path to the mine yard in time to see the last few machine operators disappear inside the entrance.

Something was wrong. Bad wrong. The yard was never this quiet. Men were always there, perched in cockpits of oversized machinery that blasted heavy metal noise against the ridge walls like the music we would love in twenty years' time. Now the yard was as silent as death. The ticking of time was noted only by the beating of my mother's heart.

She was vaguely aware of someone standing beside her now; another wife, mother, or sister. She wasn't sure. No one spoke. I didn't know if there was nothing to say or if words just wouldn't come. One minute passed. Two. Ten.

An ambulance quietly rolled into the yard. Then, another. Six vehicles in all rested between the women and the mine entrance. White hearses coated in black dirt. I don't know how their legs held the women upright. Hope and determination seemed to be strength enough. Another fifteen minutes passed. Time was an unbearable companion during hours like these.

The dark figure of the foreman slowly shuffled out of the mine, removing his helmet and tucking it beneath an arm. A white cap of hair jostled in the breeze. He ran fingers through his silky mat, leaving black zebra streaks in its wake. One of the women took an anxious step forward as the man approached. The others followed.

"Three are dead and one seriously injured," he said without preamble. "Vada Mae, you'll be accompanying Ed to the hospital when they bring him up. Edith, Sara, Betty, please follow me."

Death leaves a heartache no one can heal; love leaves a memory no one can steal.[35]

The gurney drifted across the hard-packed ground toward the ambulance; an unreal, bright day nightmare. My father's body was motionless but the blood-soaked towel on his face told the whole story. His nose had been crushed when the full weight of the crossbeam dropped on his face. A chain reaction crumbled the fragile low ceiling, burying the four miners under tons of coal.

After emergency constructive surgery and a week of intensive care, my father was released to recuperate at home. Left with a small scar in the shape of a holy cross on the bridge of his new nose, this Marine Corps veteran of Guadalcanal, Tarawa, Iwo Jima and Okinawa had learned a valuable life lesson and never went back to the mines. My parents packed up our lives, what there was of them, in the back of a '49 Ford and moved north to Chicago in search of a new career.

But for hundreds of miles around central and northern Alabama, if a man wasn't a sharecropper and he needed to work, coal mining remained the only game in town for another twenty years after that day. So, most became miners and did what life required of them. In 1952, the pay barely kept them alive topside at $18 per day, a little more than $100 for a six-day week, while women and children knew little of the true hardships these men faced beyond the mine entrance. But they all knew that guarantees down there were as fragile as the coal they picked apart. Explosions, suffocation from lack of oxygen, exposure to toxic gases and other materials, and the coal dust itself were all very real threats.

Nineteenth-century miners entered the mines equipped with the tools of their trade; picks, shovels, pry bars, breast augers, saws, axes, and

---

[35] Inscription on an Irish tombstone

tamping bars. Frequently, mine owners provided the necessary equipment, financing miners' purchases against future wages. Three-tiered dinner buckets contained their food and drinking water, and kerosene oil lamps provided dim, smoky lighting. After the turn of the twentieth century, though, carbide lamps replaced the kerosene lights. In the 1930s and 1940s, battery-powered lamps eliminated the need for an open-flame lantern.[36]

Mules pulled car loads of extracted coal along tracks and cables hoisted the cars to the mine entrance. Topside at the "tipple," each car was weighed and credited tonnage to respective miners, and "day men" dumped the coal for sorting and loading. The use of black powder and mules pulling one to two-ton capacity tram cars was standard operating procedure until the 1930s when the mines first began to "mechanize". Mechanical shakers, for example, were installed about that time to separate large lump from medium lump from nut and slack coal.

Until recently, coal extraction from the underside of ridges to deep well-holes in the Earth was a fairly basic process; undercutting, drilling, blasting, and loading. But due to growth in surface mining and improved mining technology, the amount of coal produced by one miner in one hour has more than tripled since 1978. Surface mines began using large Earth-moving equipment, such as draglines, shovels and loaders. This type of heavy machinery replaced much of the manual labor required in past operations.

Ironically, at about the same time coal markets began a severe decline in Alabama and most mines closed. The coal was still there. That wasn't the problem facing mine companies, nor was it the temporary slowdown in the markets. The operations able to continue became over constrained by newly enacted environmental protection laws, land reclamation expenses, safety regulations, and rising labor costs.[37]

---

[36] Encyclopedia of Alabama http://www.encyclopediaofalabama.org/article/h-1473
[37] Encyclopedia of Alabama http://www.encyclopediaofalabama.org/article/h-1473

Natural Resources Defense Council

NRDC works to safeguard the earth—its people, its plants and animals, and the natural systems on which all life depends.

We believe the world's children should inherit a planet that will sustain them as it has sustained us. NRDC works to ensure the rights of all people to the air, the water and the wild, and to prevent special interests from undermining public interests. NRDC does this in a number of ways.

Fighting climate change by cutting carbon pollution and expanding clean energy is the best way to build a better future for our children. NRDC is tackling the climate crisis at its source: pollution from fossil fuels. We work to reduce our dependence on these dirty sources by expanding clean energy across cities, states, and nations. We win court cases that allow the federal government to limit carbon pollution from cars and power plants. We help implement practical clean energy solutions. And we fight oil and gas projects that would pump out even more pollution.

Getting rid of toxic chemicals in our environment—in the food we eat, the air we breathe, the water we drink, and the products we buy—helps protect the health of millions of people. When the U.S. Environmental Protection Agency and U.S. Food and Drug Administration fail to protect Americans, workers, and children from dangerous chemicals, NRDC takes them to court. We fought for reforms that took millions of pounds of the most harmful pesticides off the market. And we team up with communities to wipe out indoor health hazards, including mold in public housing.

NRDC also protects other areas such as communities, energy, food, health, oceans, water, and the wilderness through litigation, science, business, partnerships and advocacy across the planet. NRDC experts ground their work in research and science to advocate for laws and policies that create lasting environmental change. NRDC experts use data and science to unearth the root causes of the problems that confront us. We use that information to blueprint transformative solutions, and we mobilize the support of partners, members, and activists to advocate for laws and policies that will protect our environment far into the future.

We combine the power of more than three million members and online activists with the expertise of some 600 scientists, lawyers, and policy advocates across the globe to ensure the rights of all people to the air, the water, and the wild. Join us at (www.NRDC.org).

On the heels of automation, came thoughts of mine safety. Unions began insisting on worker protections and companies were looking at ways to reduce costs. These two views crossed paths when management understood that fewer accidents meant an increase to the bottom line. Okay, so the real reason mining companies got interested in safety was because they had no choice. As mentioned above, the 1970s and 1980s gave us a government with a conscience that enacted environmental protections and a variety of safety laws for American workers.

One example of this legislation, The Mine Safety and Health Act of 1977 (effective March, 1978), brought a number of critical improvements to the Mine Safety and Health Act of 1969, including:

- Combination of coal, metal and non-metal mines under single legislation
- Retention of separate health and safety standards for coal mining
- Transfer of enforcement from the Department of Interior to the Department of Labor
- Renaming of the Mine Enforcement Safety Administration (MESA) to the Mine Safety and Health Administration (MSHA)
- Four annual inspections of underground coal mines
- Two annual inspections of all surface mines
- Elimination of advisory standards for metal and nonmetal mines
- Discontinuation of state enforcement plans
- Mandating of miner training
- Requirement of mine rescue teams for all underground mines
- Provision of increased involvement of miners and their representatives in health and safety activities.[38]

So, coal mines were getting a little safer for workers but there were still parts of the operation that worked against the Earth and its various species. Mining in general has severely adverse effects on the environment including loss of biodiversity, toxic methane release into the atmosphere, erosion (including sinkholes), chemical leakage, and contamination of surface water, groundwater, and soil. It took a renewed

---

[38] U.S. Department of Labor: Mine Safety and Health Administration – MSHA
https://arlweb.msha.gov/REGS/ACT/ACT1.HTM#5

interest in the environment to bring attention to additional failings of mine operations.

Before we talk about the environmental downside of mining, let's define the four major types of mining – surface, underground, placer and in-situ mining. The two methods most commonly used for coal extraction are surface and underground mining. Placer and In-situ mining are used to extract precious stones, metals (frequently uranium) and other ores.

Surface mining is used to produce most of the coal in the U.S. today because it is less expensive than underground mining. Surface mining can be used when the coal is buried less than 200 feet underground. In surface mining, giant machines remove the top-soil and layers of rock to expose large beds of coal. Once the mining is finished, the dirt and rock are returned to the pit, the topsoil is replaced, and the area is replanted. The land can then be used for croplands, wildlife habitats, recreation, or offices or stores.[39]

Underground mining, sometimes called deep mining, is used when the coal is buried several hundred feet below the surface. Some underground mines are 1,000 feet deep. To remove coal in these underground mines, miners ride elevators down deep mine shafts where they run machines that dig out the coal.[40]

---

[39] Energy Trends Insider: Coal Mining and Processing
http://www.energytrendsinsider.com/research/coal/coal-mining-and-processing/
[40] Energy Trends Insider: Coal Mining and Processing
http://www.energytrendsinsider.com/research/coal/coal-mining-and-processing/

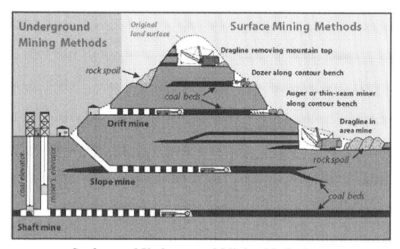

Surface and Underground Mining Methods

Placer mining is an ancient method of using water to excavate, transport, concentrate, and recover heavy minerals from alluvial or placer deposits. Examples of deposits mined by means of this technique are the gold-bearing sands and gravel that settle out from rapidly moving streams and rivers at points where the current slows down. Placer mining takes advantage of gold's high density, which causes it to sink more rapidly from moving water than the lighter siliceous materials with which it is found. Though the basic principles of placer mining have not altered since early times, methods have improved considerably.[41]

In-Situ leach mining (ISL) is a method of uranium mining, for example, where hundreds of wells are drilled in a "grid pattern" over an ore body that is located in a groundwater aquifer. Water mixed with sodium bicarbonate concentrate is pumped down into the aquifer, across the uranium ore bed, and them up and out other wells. The mining solution is injected into the aquifer under pressure in order to leach the uranium out of the ground. The leach solution strips the uranium out of the aquifer. Along with radioactive uranium, arsenic, selenium, radium and lead are

---

[41] Encyclopedia Britannica: Placer Mining
https://www.britannica.com/technology/placer-mining

also extracted.[42]

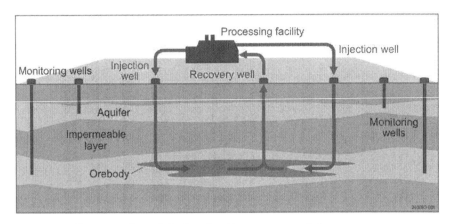

Conceptual model of an in situ recovery mining process

There are currently an estimated 1.1 trillion tons of proven coal reserves worldwide. This means that there is enough coal to last us around 150 years at current rates of production. In contrast, proven oil and gas reserves are equivalent to around 50 and 52 years at current production levels.[43]

So, what's the big deal about mining? We need energy to drive our cars and stay warm at night, don't we? We need atomic bombs to feel safe, right? We *need* gold and silver and diamonds for untold reasons. The answer is yes to all of the above questions if you want it to be. But here's the big deal. Once again, look at the coal and gas numbers above. Some of your children and most of your grandchildren will still be alive when the planet runs out of oil and not so many generations after that, the coal will be gone too. Even at present day consumption, the world will run out of dead animal fuel in the next century. The real problem, however, is the planet will not survive into the next century because of the

---

[42] Black Hills Clean Water Alliance: WHAT IS "IN SITU" MINING? https://www.excelsiormining.com/images/pdf/ExelsiorMining-InSitu-Infographic2_V4.pdf
[43] World Coal Association: Where is Coal Found? https://www.worldcoal.org/coal/where-coal-found

byproducts left after extraction of current energy sources and their long-lasting negative effects on our environment. It's time to begin a new love affair with renewable energies. Please see our chapter of *Solutions* for alternative methods of "mining" sustainable energy.

Let's look at how mining is adversely affecting our planet.

Human health risks are at the core of every adverse effect to our planet. For those in the mining business there is nothing worse than a tunnel collapse, suffocation, exposure to heavy metals or black lung disease. Lifetimes are short in the mining industry and guarantees are just dreams.

Chronic exposure to coal dust can lead to black lung disease, or pneumoconiosis, which took the lives of 10,000 miners worldwide over the last decade.[44] Rates of black lung are on the rise, and have almost doubled in the last 10 years. The U.S. National Institute for Occupational Safety and Health (NIOSH) reported that close to 9% of miners with 25 years or more experience tested positive for black lung in 2005-2006, compared with 4% in the late 1990s.[45]

But there are additional, far-reaching health concerns affecting any number of species that have never had use for heat or air conditioning or refrigerators or exterior illumination.

Despite the recent improvements in, and increasing use of renewable forms of energy, approximately 40% of the world's electricity is still generated by coal fired power stations. This figure is significantly higher in some countries, for example, approximately 70% in India and over 90% in South Africa.[46]. As long as coal is a main source of energy, the world must contend with air pollution from coal-fired power plants that include such pollutants as sulfur dioxide, nitrogen oxides, particulate

---

[44] Jeff Biggers, "What Killed the Miners? Profits Over Safety?", Huffington Post, April 6, 2010 https://www.huffpost.com/entry/who-killed-the-miners-pro_b_526602
[45] "Black lung on the rise among U.S. coal miners" World Socialist Web Site, January 11, 2010 https://www.wsws.org/en/articles/2010/01/blac-j11.html
[46] ThermoFisher Scientific https://www.thermofisher.com/blog/mining/chemical-makeup-of-fly-and-bottom-ash-varies-significantly-must-be-analyzed-before-recycled/

matter (PM), and heavy metals, leading to smog, acid rain, toxins in the environment, and numerous respiratory, cardiovascular, and cerebrovascular effects.

Specific negative health effects from coal use within the U.S. include:

- Reduction in life expectancy (particulates, sulfur dioxide, ozone, heavy metals, benzene, radionuclides, etc.)
- Respiratory hospital admissions (particulates, ozone, sulfur dioxide)
- Black lung from coal dust
- Congestive heart failure (particulates and carbon monoxide)
- Non-fatal cancer, osteoporosis, ataxia, renal dysfunction (benzene, radionuclides, heavy metals, etc.)
- Chronic bronchitis, asthma attacks, etc. (particulates, ozone)
- Loss of IQ from air and water pollution and nervous system damage (mercury)
- Degradation and soiling of buildings that can effect human health (sulfur dioxide, acid deposition, particulates)
- Global warming (CO_2, methane, nitrous oxide)
- Ecosystem loss and degradation, with negative effects on health and quality of life.[47]

Some of the main culprits of these health and ecosystem concerns are:

Methane: The damage methane release into the atmosphere has already been well documented. What you may not know is mining adds a considerable amount of methane to the air. To prevent the buildup of gases in the tunnels, methane must be constantly ventilated out of underground mines to keep miners safe. By 2020, global methane emissions from coal mines are estimated to reach nearly 800 metric tons of CO_2 equivalent (CO_2E), accounting for 9% of total global methane emissions. China leads the world in estimated coal mine methane

---

[47] ScienceWatch: Health Effects of Coal
https://www.sourcewatch.org/index.php/Health_effects_of_coal

(CMM) emissions with more than 420 metric tons of CO_2E by 2020 (more than 27 billion cubic meters annually). Other leading global emitters are the United States (84 metric tons), Russia (60 metric tons), Australia (37 metric tons), Ukraine (37 metric tons), Kazakhstan (27.5 metric tons), and India (27.4 metric tons).[48]

These numbers are scary because methane (CH_4) is the second most important greenhouse gas (GHG) after carbon dioxide (CO_2). In fact, methane is more than 25 times more potent than CO_2 on a mass basis over a 100-year time period. 64% of the methane in our air is cause by human activities. CMM represents wasted emissions to the atmosphere, while capture and use of CMM has benefits for the local and global environment.[49]

The good news: In 2015, U.S. coal mines recovered and utilized more than 33 billion cubic feet of CMM. Nearly all of this gas was sold to natural gas companies for re-sale. Globally, as of 2015, there were more than 200 operating recovery and utilization projects in about 15 countries at active or abandoned coal mines and approximately 30 more projects were in development at that time. Collectively, these projects mitigate nearly four billion cubic meters of methane each year (more than 60 million metric tons of CO_2 equivalent).[50] Still, more is needed.

Strip Mining the Geosphere: The major drawback of this type of industrial mining (also known as open cast, mountaintop or surface mining) is the damage its operations cause to the environment. Removal of large areas of topsoil can destroy entire ecosystems, and the chemicals used in mining operations can leach into the groundwater and pollute the area. Air pollution and radioactive pollution are other possible downsides.

---

[48] EPA: Coalbed Methane Outreach Program FAQs
https://www.epa.gov/cmop/frequent-questions#q6
[49] EPA: Coalbed Methane Outreach Program FAQs
https://www.epa.gov/cmop/frequent-questions#q1
[50] EPA: Coalbed Methane Outreach Program FAQs
https://www.epa.gov/cmop/frequent-questions#q8

Impacts of strip mining:

- Strip mining destroys landscapes, forests and wildlife habitats at the site of the mine when trees, plants, and topsoil are cleared from the mining area. This in turn leads to soil erosion and destruction of agricultural land.
- When rain washes the loosened top soil into streams, sediments pollute waterways. This can hurt fish and smother plant life downstream, and cause disfiguration of river channels and streams, which leads to flooding.
- There is an increased risk of chemical contamination of groundwater when minerals in upturned Earth leach into the water table, and watersheds are destroyed when disfigured land loses the water it once held.
- Strip mining causes dust and noise pollution when topsoil is disrupted with heavy machinery and coal dust is created in mines.

Environmental hazards are present during every step of the open-pit mining process. Hard rock mining exposes rock that has lain unexposed for geological eons. When crushed, these rocks expose radioactive elements, asbestos-like minerals, and metallic dust.[51] The result of all this is barren land that stays contaminated long after a mine shuts down, including death of flora and fauna, and erosion of land and habitat much larger than the mine site itself.

Although the U.S. and many countries require reclamation plans for coal mining sites, undoing all the environmental damages to water supplies, destroyed habitats, and poor air quality is a long and problematic task. This land disturbance is on a vast scale. Between 1930 and 2000, U.S. coal mining activities altered about 2.4 million hectares [5.9 million acres] of natural landscape, most of it originally forest. Attempts to re-seed land destroyed by coal mining are difficult because the mining process has so thoroughly damaged the soil. For example, in Montana,

---

[51] Mission 2016: The Future of Strategic Natural Resources
https://web.mit.edu/12.000/www/m2016/finalwebsite/problems/mining.html

replanting projects had a success rate of only 20-30%, while in some places in Colorado only 10% of oak aspen seedlings that were planted survived. The overall restoration rate (the ratio of reclaimed land area to the total degraded land area) of mine wasteland was only about 10–12%.[52]

Geological Hazards: Quite often, mining involves drilling into or blowing up land. This creates significant physical disturbances to the ecosystems at a site. What's more, the features of the land from before the area was mined can often not be replaced or replicated.[53] Earth collapses, landslides, cave-ins and sinkholes are all forms of geological instability caused by mining activities. All mines are temporary structures. They can remain active for many years, but they will eventually run out of minerals and cease operations. This does not automatically mean that the environment and wildlife will no longer suffer.

Whether resulting from surface or underground mining failure to backfill a depleted mine can lead to a problem called subsidence, which occurs when abandoned mines collapse. This will undo any efforts to reestablish a healthy ecosystem in the area, and often render it useless for many years to come. The problem only increases if contaminants were left on the site, since removing them after a collapse is exceedingly difficult.[54]

One such geological instability occurred in 1980. A salt mine underneath Lake Peigneur in Louisiana collapsed when an oil drilling operation breached part of the mine, and the resulting collapse sucked in barges, vehicles and buildings. Once the lake settled, the entire shoreline and

---

[52] Environment: Effects Of Mining on the Environment and Human Health
https://www.environment.co.za/mining/effects-of-mining.html
[53] Greentumble: How Does Mining Affect the Environment?
https://greentumble.com/how-does-mining-affect-the-environment/
[54] Pegasus Foundation: Effects of Mining on the Environment and Wildlife
https://www.pegasusfoundation.org/effects-of-mining-environment-wildlife/

region had been drastically changed by the collapse.[55]

In-Situ Leach Mining (ISL): This type of mining is just plain unforgivable for what it does to Mother Earth and every species that ever inhabited her. Drilling must be done directly in a water-bearing aquifer. Yep, from the description above in situ mining (literally meaning "in position") burrows into a water-bearing aquifer to leach tiny bits of uranium and other heavy metals to forever violate our drinking water. Water at an in situ leach uranium mine has never been returned to its original condition. Pollutants that have been left in the water at in situ leach uranium mines after "restoration" include toxic heavy metals and radioactive materials. Just one proposed mine — the Powertech Uranium project near Edgemont (South Dakota) — would consume over two and a half billion gallons of water during its lifetime. This is according to the company's own figures.[56]

There is another impact to our health we should mention here because it really doesn't get enough media coverage. **Mine tailings** are the materials left over after the process of separating the small amount of valuable ore from the bulk of the uneconomic debris. Generally, there are two types of tailings; hard rock and wet. Hard rock tailings are sometimes pushed aside or reburied in and on the mine when all of the ore has been mined from the site. However, it contain naturally occurring contaminants, such as arsenic, cadmium, lead, mercury, and selenium that can be leached from precipitation and then contaminate groundwater.

When tailings end up in the form of sludge or a wet mud substance, wet storage often requires long-term oversight, to monitor and attempt to mitigate contaminated groundwater movement, and to maintain any crucial facilities, such as dams. Many wet storage facilities require perpetual water treatment.

---

[55] Reference.com: What Are the Disadvantages of Mining? https://www.reference.com/science/disadvantages-mining-7db6f3b8c2277cf8
[56] Black Hills Clear Water Alliance https://bhcleanwateralliance.org/what-is-in-situ-mining/

Disposal of mine tailings is usually the single biggest environmental concern facing a hard rock metal mine, and creates very long-term environmental liabilities which future generations must manage. Many mine tailings do not become appreciably safer over time, even if stored properly, and therefore must be stored for an indefinite period using current technology.

The historically-used alternative to storage was to dispose of tailings in the most convenient way possible (such as river dumping), which led to widespread environmental contamination in mining areas. This was nominally viable in earlier eras, but human production of mine tailings has increased by several orders of magnitude in the modern age, making such methods unacceptable in today's societies.

Many modern hard rock metal mines dispose of tailings as a wet mud, held in pits lined with clay or a synthetic liner. Many other mines put the tailings back into the original mining pit. Some large mines use entire existing valleys sealed off with Earthen dams, and others store tailings in natural lakes. In most cases, disposal pits are covered with water, forming an artificial lake which reduces the rate of acid formation.[57]

Sometimes these containers and dams fail. Two of the worst disasters occurred in Brazil in 2015 and 2019. Only 1,177 days separate accidents associated with the Fundão ore reject dams in Mariana and the Córrego do Feijão mine in Brumadinho in the metropolitan region of Belo Horizonte, both in southeastern Brazil. In the first incident, in November 2015, the toxic sludge expelled by the structure killed 19 people, buried villages, left thousands of people homeless, and reached the sea. At the time it was considered one of the country's biggest socio-environmental disasters in the mining sector.

And then on January 25th, 2019, about 78 miles from Mariana, another tragedy struck in the state of Minas Gerais. The full impact of the Brumadinho accident is still being evaluated, but at least 65 people have

---

[57] Ground Truth Trekking: Mine Tailings
http://www.groundtruthtrekking.org/Issues/MetalsMining/MineTailings.html

been reported dead, victims of the tailings mud stored at Dam I of the Córrego do Feijão Mine, and about 280 were missing at the time of this writing (February 1, 2019).

Environmental experts say that the collapse of those two dams, operated respectively by the Samarco (joint-venture of BHP Billinton and Vale S.A) and Vale, could have been avoided. Stricter licensing laws and state oversight and the adoption of more modern technology could transform the Brazilian mining sector, making such incidents less likely, experts told National Geographic Brasil.[58]

Turning from extraction to the burning of decayed plant matter, several principal and deadly emissions resulting from coal combustion:

- Sulfur dioxide (SO2), which contributes to respiratory illnesses and acid rain. Sulfur dioxide, associated SOx, and secondary pollutants can contribute to respiratory illness by making breathing more difficult, especially for children, the elderly, and those with pre-existing conditions. Longer exposures can aggravate existing heart and lung conditions, as well. Sulfur dioxide and other SOx are partly culpable in the formation of thick haze and smog, which can impair visibility in addition to impacting health.

    Beyond human health impacts, sulfur dioxide's contribution to acid rain can cause direct harm to trees and plants by damaging exposed tissues and, subsequently, decreasing plant growth. Other sensitive ecosystems and waterways are also impacted by acid rain.[59]

- Nitrogen oxides (NO_X), which contribute to respiratory illnesses and smog. Nitrogen is the most common part of the air we breathe: about 80% of the air is nitrogen. When air is heated, like in coal boilers, nitrogen atoms break apart and join with oxygen, forming

---

[58] National Geographic: Brazil's deadly dam disaster may have been preventable
https://www.nationalgeographic.com/environment/2019/01/brazil-brumadinho-mine-tailings-dam-disaster-could-have-been-avoided-say-environmentalists/
[59] Minnesota Pollution Control Agency: Sulfur dioxide (SO2)
https://www.pca.state.mn.us/air/sulfur-dioxide-so2

nitrogen oxides (NO$_X$) (rhymes with "socks"). NOx can also be formed from the atoms of nitrogen that are trapped inside coal.

In the air, NOx is a pollutant. Coal combustion releases oxides of nitrogen, which react with volatile organic compounds in the presence of sunlight to produce ground-level ozone, the primary ingredient in smog. Asthma exacerbations have been linked specifically to exposure to ozone. Nitrogen oxide also contributes to fine particulate matter (PM), found in soot, which is also linked to a host of serious health effects.

Exposures to ozone and PM are both correlated with the development of and mortality from lung cancer. Recent research suggests that nitrogen oxides and PM$_{2.5}$ (atmospheric particulate matter that have a diameter of less than 2.5 micrometers. Particles in this category are so small that they can only be detected with an electron microscope.), along with other pollutants, are associated with hospital admissions for potentially fatal cardiac rhythm disturbances. Cities with high nitrogen dioxide (NO$_2$) concentrations have death rates four times higher than those with low NO$_2$ concentrations, suggesting a potential correlation. NOx also harms the environment, contributing to acidification of lakes and streams (acid rain). Aging coal plants "grandfathered" in after passage of the Clean Air Act have been particularly linked to large quantities of harmful emissions.[60]

- Particulates, which contribute to smog, haze, and respiratory illnesses and lung disease. The PM level, individually and in combination with NO2 in air, increases the concentration of free radical based reactive oxygen species (ROS) and contributes to DNA mutation, and damage of protein and lipids which may constitutively activate membrane proteins which leads to the development of some serious diseases, including lung cancer,

---

[60] SourceWatch.org: Nitrogen Oxide
https://www.sourcewatch.org/index.php/Nitrogen_oxide

cardiovascular diseases and reproductive disorders. The interaction of PM with DNA leads to the formation of DNA adducts impairing neurodevelopment, intelligence quotient (IQ) levels and intelligence in children.[61]

- CO_2, which is the primary greenhouse gas produced from burning fossil fuels (coal, oil, and natural gas). That's a huge problem for the climate because more new coal-fired power plants have been built worldwide in the past decade than in any previous decade, with no sign of slowing down.

Those existing coal and other fossil fuel-fired power plants emit billions of tons of CO_2 each year and account for about 26% of global greenhouse gas emissions — double that of the transportation sector. In the U.S. alone, burning coal emitted 1.87 billion tons of CO_2 in 2011, according to the U.S. Energy Information Administration. Worldwide, coal-burning released 14.4 billion tons of CO_2 in 2011.

But a study extends those emissions out to the full lifespan of each of the existing power plants — 40 years per plant — and estimates that together they will spew out 300 billion tons of CO_2 before they are retired, up from 200 billion tons of CO_2 emissions that were committed from the power plants that existed in 2000, the study says. In other words, the power plants operating today are committed to emitting 300 billion tons of CO_2 in the future, enough to contribute an additional 20 ppm of CO_2 to the atmosphere globally, Princeton University professor emeritus of mechanical and aerospace engineering and study co-author Robert Socolow told Climate Central.[62]

---

[61] ScienceDirect: Human health and environmental impacts of coal combustion and post-combustion wastes
https://www.sciencedirect.com/science/article/pii/S2300396017300551
[62] Climate Central: Coal Plants Lock in 300 Billion Tons of CO2 Emissions
https://www.climatecentral.org/news/coal-plants-lock-in-300-billion-tons-of-co2-emissions-17950

- Mercury and other heavy metals, which have been linked to both neurological and developmental damage in humans and other animals. The presence of high quantities of arsenic, copper, and selenium in fly ash (Zhang, 2014) indicate adverse impacts to both water and soil (Nriagu & Pacyna, 1988). Generally, about 90% of coal ash is comprised of iron, aluminum, silicon, and calcium in their oxide form. Sodium, magnesium, potassium, titanium are minor constituents, representing about 8% of the mineral matter component, although some trace matters such as arsenic, cadmium, lead, mercury, and selenium are also present and represent up to 1% of the total ash composition (EPRI, 2009). These trace element wastes, upon the dumping of ash in selected land and ponds, cause serious environmental problems, such as leachate (Nalawade et al., 2012, Lokeshappa et al., 2010). Leachate is the liquid formed when permeable material (either dissolved or suspended) percolates with water (Tiwari, Bajpai, Dewangan, & Tamrakar, 2015). Volatilization, melting, decomposition and oxidation are key mechanisms, which release and transport the trace metals from coal fly ash into the environment of soil and water, for contaminating the surface and groundwater (Lokeshappa and Dikshit, 2012, Kim et al., 2003). Water creatures, including fish, intake trace metal pollutants through food, skin and gills. Transfer of these pollutants to the bloodstream culminates in the bioaccumulation of trace metals in liver, gills or kidney. Such bioaccumulation of toxic metals in the food web leads to adverse impacts on both human health and the environment (Akintujoye et al., 2013, Al-Kahtani, 2009). Similarly, the accumulation of heavy metals of fly ash on the scales of fish induces excessive damage and may culminate in the blockage of scale formation (Shikha & Sushma, 2011).[63]

- Few people have heard of coal ash (better known as fly ash, bottom ash, and boiler slag), which is created in a coal-fired power plant. It

---

[63] ScienceDirect: Human health and environmental impacts of coal combustion and post-combustion wastes
https://www.sciencedirect.com/science/article/pii/S2300396017300551

contains arsenic, mercury, lead, and many other heavy metals. Not something you'd want to sprinkle on a salad. But U.S. fossil fuel plants produce 140 million tons of the stuff every year in the process of combusting coal, making it the nation's second-largest waste stream behind household trash. Every now and then we get a reminder of just how much coal ash is around. In 2008, a coal ash pond broke through its dam in Kingston, Tennessee, covering hundreds of acres in toxic sludge. In early 2014, a spill released more than 30,000 tons of coal ash into North Carolina's Dan River. The U.S. EPA finally got around to regulating the waste, but the thinly veiled rules are a major disappointment to environmentalists and the millions of Americans living near coal ash pits and ponds.[64]

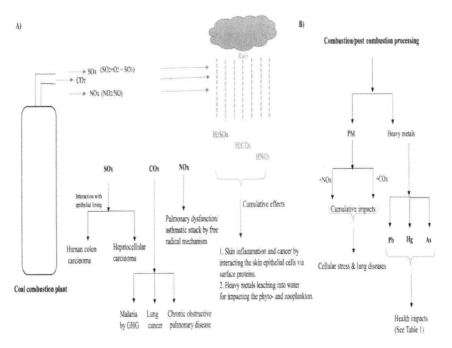

Science Direct: Journal of Sustainable Mining

Coal ash can have a different chemical makeup depending on where the

---

[64] NRDC: Coal Ash, Fly Ash, Bottom Ash, and Boiler Slag
https://www.nrdc.org/onearth/coal-ash-fly-ash-bottom-ash-and-boiler-slag

coal was mined. Broadly speaking, coal ash is a pollutant, and it contains acidic, toxic, and radioactive matter. The EPA found that significant exposure to bottom ash and other components of coal ash increases a person's risk of developing cancer and other respiratory diseases. In addition, storage lagoons can pollute groundwater, and if ingested, the arsenic contaminated water increases a person's risk of developing cancer. Inhalation is not as much of an issue with bottom ash as it is much heavier than fly ash. However, ingestion of bottom ash can have nervous system impacts, cognitive defects, developmental delays, and behavioral problems. Moreover, it increases a person's chance of developing lung disease, kidney disease, and gastrointestinal illness.[65]

Even though most of this chapter focused on coal mining, I want to remind you that most of what is said here also applies to other forms of mining too. Diamond, gold, silver, uranium and other ore mining are just as harmful to our world.

Even though diamond mining uses far fewer chemical contaminants in processing, the process still has a devastating impact on the Earth's environment and its many species. A century of reckless diamond mining has taken a heavy toll on Angola's environment. Irresponsible diamond mining has caused soil erosion, led to deforestation, and forced local populations to relocate. Angola's diamond industry has been particularly careless in protecting rivers and streams from exploitation. Diamond miners have re-routed rivers and constructed dams to expose riverbeds for mining, with disastrous effects on fish and wildlife.

In extreme cases, diamond mining can cause entire ecosystems to collapse. Diamond miners in the Kono district of eastern Sierra Leone have left behind thousands of abandoned mining pits. Wildlife has vanished, topsoil has eroded, and land once suitable for farming is now a desolate moonscape. The mining pits have created a public health disaster as well. When the pits fill with stagnant rainwater, they become

---

[65] PSR. (July 17, 2015). Coal Ash: Hazardous to Human Health [Online] http://www.psr.org/assets/pdfs/coal-ash-hazardous-to-human-health.pdf

infested with mosquitoes, spreading malaria and other water-borne diseases.[66]

In addition, diamond mining continues to use practices that exploit workers, children, and communities. A million diamond diggers in Africa earn less than a dollar a day. Miners are dying in accidents, child labor is widespread, and corrupt leaders are depriving diamond mining communities of funds badly needed for economic development.[67]

In August 2019, news of the amazon rainforest burning stunned the world. Many scientists believe cattle ranchers clearing land caused the flames, spurring groups around the world — including the government of Finland — to call for a boycott of Brazilian beef. But to boycott all of the products damaging the Amazon, you'd have to do much more than give up steak. You'd have to toss out your phone, laptop, wedding band, and anything else with gold in it.

Tiny electrical currents are constantly running through your iPhone, Alexa speaker, and laptop — and carrying those currents is gold, a fantastic conductor of electricity that's also resistant to corrosion. While there isn't much gold inside a single device — an iPhone 6, for example, contains 0.014 grams, or around 50 cents' worth — in the aggregate, the amount is staggering. According to market researcher Gartner, over 1.5 billion smartphones were sold last year, with 1.3 billion of them being Android devices. It was followed by 215 million iOS devices.

"There's no way to get the gold out without destroying the forest. The more acres you cut down, the more gold you get. It's directly proportional," Miles Silman, the cofounder of Wake Forest University's Center for Amazonian Scientific Innovation (CINCIA), told BuzzFeed

---

[66] Brilliant Earth: Environment https://www.brilliantearth.com/blood-diamond-environmental-impact/
[67] Brilliant Earth: Labor & Community https://www.brilliantearth.com/conflict-diamond-child-labor/

News.[68]

On the other hand, mining of metal and precious ores (please also refer back to In Situ mining above) presents very different concerns for the environment. Operations and waste products associated with metal extraction and processing are the principal causes of environmental concerns about metal mining. The general issues are:[69]

- **Physical disturbances to the landscape** – Like coal mining, since the 1970s other forms of mining have also gravitated to surface mining. The amount of overburden and waste rock in open pit mines is significantly greater than the ore produced. Tremendous volumes of waste rock are removed from the pits and deposited in areas nearby.

  Waste piles from processing, such as tailings impoundments, leach piles, and slag piles vary in size, but can be very large. The impoundments associated with some of the largest mills, such as at open pit copper mines, can cover thousands of acres and can be several hundred feet thick. Heap leach piles can cover tens to hundreds of acres and be a few hundred feet high. They resemble waste rock piles in location and size, but are more precisely engineered. Slag is a glassy by-product of smelting; slag piles can cover tens to hundreds of acres and be over 100 hundred feet high. The same biodiversity and habitat loss occur with ore mining.

- **Soil and water contamination from waste rock and tailings** – Although the character of waste rock varies with the type of ore, many waste rocks contain sulfide minerals associated with metals, such as lead, zinc, copper, silver, or cadmium. An important sulfide mineral common in waste rock is pyrite (iron sulfide, aka fool's

---

[68] Buzzfeed.News: It's Not Just Fires. Your Phone Is Also Destroying The Amazon. https://www.buzzfeednews.com/article/nicolenguyen/gold-mining-amazon-rainforest
[69] American Geosciences Institute: How can metal mining impact the environment? https://www.americangeosciences.org/critical-issues/faq/how-can-metal-mining-impact-environment

gold). When pyrite is exposed to air and water, it undergoes a chemical reaction called "oxidation." The oxidation process produces acidic conditions that can inhibit plant growth at the surface of a waste pile. Bare, non-vegetated, orange-colored surface materials make some waste rock areas highly visible, and they are the most obvious result of these acidic conditions.

Where acid rock drainage occurs, the dissolution and subsequent mobilization of metals into surface water bodies and groundwater is probably the most significant environmental impact associated with metallic sulfide mineral mining. Acidic and metal-bearing groundwater occurs in abandoned underground mine workings and deeper surface excavations that encounter the groundwater of a mineralized area. Because they are usually located at or below the water table, underground mines act as a type of well which keeps filling with water. Because these waters migrate through underground mine workings before discharging, they interact with the minerals and rocks exposed in the mine. If sulfide minerals are present in these rocks, especially pyrite, the sulfides can oxidize and cause acid rock drainage.

Tailings produced from the milling of sulfide ores — primarily copper, lead, and zinc ores — may have concentrations of pyrite that are greater than those common in waste rock. Also, because tailings are composed of small mineral particles the size of fine sand and smaller, they can react with air and water more readily than waste rocks. Therefore, the potential to develop acidic conditions in pyrite-rich tailings is very high.

- **Air contamination** - sulfur dioxide has been the most common emission of concern, because it reacts with atmospheric water vapor to form sulfuric acid or "acid rain." The acidic conditions that develop in the soils where these emissions precipitate can harm existing vegetation and prevent new vegetation from growing. Barren areas near smelting operations have been an enduring environmental impact of historical smelting.

For the conclusion to our discussion of worldwide mining, I'll leave you with this sobering summation of mining's part in damaging our planet and all life on her. The negative effects of mining can no longer be disputed:

- **Destruction of Landscapes and Habitats:** Strip mining also known as surface mining, involves the stripping away of Earth and rocks to reach the coal underneath. If a mountain happens to be standing in the way of a coal seam within, it will be blasted or levelled - effectively leaving a scarred landscape and disturbing ecosystems and wildlife habitat.

- **Deforestation and Erosion:** As part of the process of clearing the way for a coal mine, trees are cut down or burned, plants uprooted and the topsoil scraped away. This results in the destruction of the land (it can no longer be used for planting crops) and soil erosion. The loosened topsoil can be washed down by rains and the sediments get into rivers, streams and waterways. Downstream, they can kill the fish and plant life and block river channels which cause flooding.

- **Groundwater Contamination:** The minerals from the disturbed Earth can seep into groundwater and contaminate water ways with chemicals that are hazardous to our health. An example would be Acid Mine Drainage. Acidic water can flow out of abandoned coal mines. Mining has exposed rocks which contain the sulfur-bearing mineral, Pyrite. This mineral reacts to air and water to form sulfuric acid. When it rains, the diluted acid gets into rivers and streams and can even seep into underground sources of water.

- **Chemical, Air & Dust Pollution:** Underground mining allows coal companies to dig for coal deeper into the ground. The problem is that huge amounts of Earth and rock are brought up from the bowels of the Earth. These mining wastes can become toxic when they are exposed to air and water. Examples of toxins are mercury, arsenic, fluorine and selenium. The amount of dust generated in mining operations can be carried to nearby towns by the wind. These dust particles can cause all kinds of health problems for humans who are exposed to it.

- **Methane in the Atmosphere:** Coal mine methane emissions from underground mining are often caught and used as town fuel, chemical feedstock, vehicle fuel and industrial fuel – but very rarely is everything captured. Methane is less prevalent in the atmosphere as compared to CO_2, but it is 20 times more powerful as a greenhouse gas.

- **Coal Fires:** Fires from underground mines can burn for centuries! These fires release smoke into the atmosphere - smoke which contains CO_2, carbon monoxide (CO), methane, nitrous oxide (NOx), sulfur dioxide (SO2) and other toxic greenhouse gases.

- **Health Hazards:** Coal dust inhalation can cause black lung disease. Miners and those who live in nearby towns are the most affected. Cardiopulmonary disease, hypertension, COPD, and kidney disease are found in higher than normal rates in people who live near coal mines.

- **Displacement of Communities:** All of these negative effects force people to move to other places as their air and water gets polluted and expanding coal mines make use of more and more of their habitat.[70]

The Cost of Inaction

If no action is taken to mitigate the many environmental problems inherent to modern mining, the end cost for governments and communities would be devastating. As of 2011, mines in China released 9,600 to 12,000 cubic meters of toxic gas containing flue dust concentrate, hydrofluoric acid, sulfur dioxide, and sulfuric acid for each

---

[70] The World Counts: Cheap But Dirty Fuel
https://www.theworldcounts.com/stories/Negative-Effects-of-Coal-Mining

ton of rare Earth elements produced. Additionally, nearly 75 cubic meters of acidic waste water and one ton of radioactive waste residue have been generated (Paul & Campbell, 2011).[71]

China and the United States, the world's two largest emitters of CO_2, will incur some of the highest social costs of carbon of all countries, the scientists report, with respective estimated impacts of US$24 per ton and $48 per ton of emitted CO_2. India, Saudi Arabia and Brazil are also high emitters. In these countries — unlike in Canada, northern Europe and Russia — temperatures are already above the economic optimum. And climate-induced damage increases with wealth and economic growth, meaning that more-valuable property might sit in harm's way.

Combined country-level costs (and benefits) add up to a global median of more than $400 (USD) in social costs per ton of CO_2 emitted — more than twice previous estimates. On the basis of CO_2 emissions in 2017, that's a global impact of more than $16 trillion. The new analysis is based on a set of climate simulations, rather than a single climate model, and the authors calculated future harm using empirical damage functions that were independently developed for that purpose.[72]

There is a cost of inaction and there is also a cost of dying without dignity. Today in the United States, this documentary on assisted suicide by climate change we are currently experiencing is running 16X speed. There is no way to slow it down to normal speed or change its inevitable end. But we are in this together and there is a way out. Please see our chapter on *Solutions* for proven methods to begin living more sustainable lives.

---

[71] Mission 2016: The Future of Strategic Natural Resources
https://web.mit.edu/12.000/www/m2016/finalwebsite/problems/mining.html
[72] Nature International Journal of Science: The costs of climate inaction
https://www.nature.com/articles/d41586-018-06827-x

Chapter 3: Hydrosphere/Hydrologic Cycle

Ahhh...water, water, water... Is there anything more glorious than soaking in a hot bath when you're cold or diving into a cool pool when you're hot? I think it must soothe us inside our deep-seated memory of floating safe and protected in the innocence of the womb. In many ways, relaxing in the natural waters on the surface of our planet is truly returning to the womb.

If all water on the planet (hydrosphere) was our blood, the hydrologic cycle would be our circulatory system. The hydrosphere includes lakes, swamps, rivers, seas, and sometimes includes water *above* the Earth's surface, such as clouds. This water moves through the hydrologic cycle from one place to another in delicate, yet eternal processes of mist, steam, clouds, rain, snowfall, streams to rivers to seas and so on. We will discuss human impacts on the hydrosphere at length shortly, but for now let's get a firm grasp on what it is and how it works.

Drop of Water

The Earth processes and retains water with machine-like precision via the hydrologic cycle. This is an astonishing process by which water from our bodies of water vaporizes into millions of tiny particles which form into clouds. Generally, as clouds lose their water via rain or snow, it falls to the Earth. But as we will see, we are in the midst of a critical water shortage.

Wait a minute...we are a biosphere, right? I don't know about you but I've had difficulty processing the logic of how we are "running out" of water. I recently read an eye-opening National Geographic article about the 'Clean Water Crisis' and realized, I had never considered the concept that fresh water is a finite resource on this planet. For some reason I thought water was a byproduct of some mysterious component of the natural functioning of Earth cycles. I mean, we are almost a water planet. But, what is here has always been here and soon there won't be enough.

Although 70% of the surface of the Earth is covered by water, only 2.3% is fresh water. What I learned in *A Clean Water Crisis*[73] is less than 1% of that fresh water is accessible. Okay, so that means, there is a limited amount of freshwater sources generated in that water cycle that humans can access. The United Nations reports that water use has grown at more than twice the rate of population increase in the last century and those numbers will only continue to rise:

By 2025, an estimated 1.8 billion people will live in areas plagued by water scarcity, with two-thirds of the world's population living in water-stressed regions as a result of use, growth, and climate change. The challenge we now face as we head into the future is how to effectively conserve, manage, and distribute the water we have.[74]

Now I understand; the water cycle hasn't changed, it's the rate of consumption of fresh water vs. available sources that has changed. The virus human doing what viruses do; multiplying like wildfire, assimilating all the resources it can consume. My eco-Shero Sylvia E. Earle, the former chief scientist of the National Oceanic and Atmospheric Administration (NOAA) and National Geographic's lead oceanographer says, "No blue; no green." Meaning, if there is no water, there is no life. All life depends on maintaining and preserving our precious water resources.

We're all kids at heart, but the children of this generation have a tremendous burden to bear as Earthlings. In many ways, their childhoods

---

[73] National Geographic: Freshwater Crisis
https://www.nationalgeographic.com/environment/freshwater/freshwater-crisis/
[74] United Nations Foundation: Connecting you to the UN, Collaborating for Impact.
https://unfoundation.org/

are being severely tapered by the precarious state of our planet. But there are noble scientists out there creating publications to explain the very complicated issues of our beloved planet in ways that make them approachable and…well, not so scary. I'm borrowing this graphic from the U.S. Geological Survey[75]:

---

[75] Science Kids: Weather Facts
http://www.sciencekids.co.nz/sciencefacts/weather/thewatercycle.html

THE ENDANGERED EARTHLINGS' HANDBOOK

The Water Cycle

Water

Before we dive into our examination of water, (buh dum ching) here are a few terms to clarify[76]:

- Potable water – Drinkable water
- Groundwater - Water that collects or flows beneath the Earth's surface, filling the porous spaces in soil, sediment, and rocks.
- Aquifer - Any geological formation containing or conducting groundwater, especially one that supplies water for wells, springs, etc.
- River basin - A river basin is the portion of land drained by a river and its tributaries. A river basin comes closer than any other defined area of land, with the exception of an isolated island, to meeting the definition of an ecosystem in which all things, living and non-living, are connected and interdependent.[77]
- Watershed - An area of land that contains a set of streams and rivers that all drain into a larger body of water, for example, a large river, lake or an ocean.[78] This also includes flow below ground (i.e. sub-flow).
- Riparian - Related to the banks of rivers and streams and to wetlands.
- Littoral - Relating to or situated on the shore of a sea or lake.
- Nonpoint source pollution - Nonpoint source pollution (NPS) generally results from land runoff, precipitation, atmospheric deposition, drainage, seepage or hydrologic modification. NPS pollution, unlike pollution from industrial and sewage treatment plants, comes from many diffuse sources. NPS pollution is caused by rainfall or snowmelt moving over and through the ground. As the runoff moves, it picks up and carries away natural and human-

---

[76] Dictionary.com: Water Terms https://www.dictionary.com/
[77] Milwaukee Riverkeeper Apr 13, 2015, What's a River Basin? What's a Watershed?
[78] Rivers and Streams: What is a watershed? http://www.mbgnet.net/fresh/rivers/shed.htm

THE ENDANGERED EARTHLINGS' HANDBOOK

made pollutants, finally depositing them into lakes, rivers, wetlands, coastal waters and groundwater.[79]

- Point source pollution - Point source pollution refers to a single identifiable source of air, soil, water, groundwater, thermal, noise or light pollution.

When you consider the fact that nearly every human resides on some form of a watershed, imagine each contaminant we use daily; motor oil, pesticides, even the bleach you just used to clean your patio…it all washes down the watershed pathways and can eventually makes its way to your faucet or the sea. Every source of water on this planet, fresh or other, is a unique ecosystems on which millions of species, not the least of which is humans, are completely dependent.

In May, 2019, a study by the Environmental Working Group and Northeastern University found that nearly every American in the country is exposed to unhealthy drinking water. In 43 states, drinking water sites are contaminated with PFAS (Per- and polyfluoroalkyl substances are a group of man-made chemicals that includes PFOA, PFOS, GenX, and many other chemicals… chemicals are very persistent in the environment and in the human body – meaning they don't break down and they can accumulate over time.[80]) and PFOA (Perfluorooctanoic acid -conjugate base perfluorooctanoate—also known as C8—is a perfluorinated carboxylic acid produced and used worldwide as an industrial surfactant in chemical processes and as a material feedstock, and is a health concern and subject to regulatory action and voluntary industrial phase-out.[81]).

Many of these chemicals are known to cause:

- Birth defects

---

[79] EPA: What is nonpoint source pollution?
https://19january2017snapshot.epa.gov/nps/what-nonpoint-source_.html
[80] EPA - Basic Information on PFAS: [80] https://www.epa.gov/pfas/basic-information-pfas
[81] Wikipedia: Perfluorooctanoic acid
https://en.wikipedia.org/wiki/Perfluorooctanoic_acid

- Cancers
- Infertility
- Child development
- Immune deficiency[82,83]

Meet Buck Bailey, perhaps the most well know victim of PFOA, or C8 contamination.

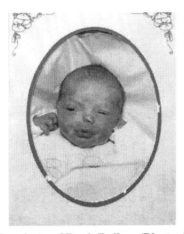

A baby photo of Buck Bailey. (Photo: Emily Kassie/Huffington Post Highline)

Buck Bailey (Photo: Emily Kassie/Huffington Post Highline)

---

[82] CBS NEWS: New study claims 43 states expose millions to dangerous chemical in drinking water https://www.cbsnews.com/news/drinking-water-may-contain-pfas-chemicals-in-43-states-according-to-new-study-by-environmental-working-group/
[83] ATSDR: What are the health effects? https://www.atsdr.cdc.gov/pfas/health-effects.html

The report also mentions data from the Pentagon showing that an estimated 19 million people are exposed to PFAS/PFOA contaminated water, in at least 610 locations.

On August 16, 2019, a steel factory in northwest Indiana (ArcelorMittal) released an apology for dumping toxic levels of cyanide and ammonia-nitrogen in the Little Calumet River. The incident killed loads of fish and shut down the Indiana Dunes National Park, a popular summer recreation spot. Although the Indiana Department of Environmental Management learned on a Monday about the steel mill's spill, they failed to inform the city until Thursday.[84]

Besides being, apparently, the Nazi's preferred method of self-dispatch in times of war; cyanide is a deadly substance and is no friend to the environment either. In this day and age, knowing what we know about the delicate status of our planet, the demand for extractive materials, (minerals, gems, mineral fuels and coal) is increasing.[85] Our executive director Paul Hollis explained the effects of mining in depth in the previous chapter. Mine tailings are the ore waste of mines, and are typically a mud-like material. Worldwide, the storage and handling of tailings is a major environmental issue. Many tailings are toxic and must be kept perpetually isolated from the environment[86]. Cyanide "heap leaching" is used for very low-quality ore, or sometimes to reprocess waste material from other extraction methods (e.g. leftover mine "tailings"). A large outdoors mound of ore is sprayed with a cyanide solution that drips through the rock over time (leaching out the remaining

---

[84] WGN 9: Officials delayed notifying Indiana city of chemical spill, mayor says
https://wgntv.com/2019/08/16/cyanide-spill-into-little-calumet-river-kills-fish-closes-some-indiana-beaches/

[85] University of Exeter: Mine tailings dams: Characteristics, failure, environmental impacts, and remediation
https://ore.exeter.ac.uk/repository/bitstream/handle/10871/34385/Kossoff_etal_Oct2014_AG_manuscript_accepted.pdf

[86] Ground Truth Trekking: Mine Tailings
http://www.groundtruthtrekking.org/Issues/MetalsMining/MineTailings.html

extractive minerals and into (often unlined) ponds for future mineral extraction from the solution. The cyanide and mineral solution can subsequently contaminate surface water bodies, as well as groundwater (including the soil)).[87]

Our sheer water consumption alone puts our habits at the top of the water chain. Feeling full? It might be because the average American consumes 32,911 glasses of water per day. Okay, not in your gullet. 96% of our individual water footprint (or 31,595 glasses) comes from ranching, agriculture and manufacturing, so if you eat or wear shoes, you're a culprit.[88]

Another factor limiting the availability of drinkable water, besides pollution and contamination, is the ever-increasing temperatures on our planet that are causing massive and widespread drought. In the U.S., we have a program in place to insure the ongoing preservation of our precious resources by watching, very closely, what's happening to our environment. The Global Change Research Act of 1990 mandates that the U.S. Global Change Research Program (USGCRP) deliver a report to Congress and the President no less than every four years that "1) integrates, evaluates, and interprets the findings of the Program…; 2) analyzes the effects of global change on the natural environment, agriculture, energy production and use, land and water resources, transportation, human health and welfare, human social systems, and biological diversity; and 3) analyzes current trends in global change, both human-induced and natural.[89]

On November 23, 2018, the U.S. government released the Fourth National Climate Assessment (NCA4.) We will make reference to this document several times throughout the book, but with regard to the U.S.

---

[87] Ground Truth Trekking: Gold Cyanidation
http://www.groundtruthtrekking.org/Issues/MetalsMining/GoldCyanidation.html
[88] You Actually Use 32,911 Glasses of Water Per Day (and The Scary Thing That Means)
https://theheartysoul.com/actually-use-32911-glasses-water-per-day-scary-thing-means/
[89] U.S. Global Change Research Program - Climate Science Special Report
https://www.nrc.gov/docs/ML1900/ML19008A410.pdf

CLEAN WATER CHANGES EVERYTHING

663 million people in the world live without clean water. That's nearly 1 in 10 people worldwide. Or, twice the population of the United States. The majority live in isolated rural areas and spend hours every day walking to collect water for their family. Not only does walking for water keep children out of school or take up time that parents could be using to earn money, but the water often carries diseases that can make everyone sick.

Access to clean water means education, income and health - especially for women and kids. There are four significant areas where access to clean water is most critical:

HEALTH - Diseases from dirty water kill more people every year than all forms of violence, including war. 43% of those deaths are children under five years old. Access to clean water and basic sanitation can save around 16,000 lives every week.

TIME - In Africa alone, women spend 40 billion hours a year walking for water. Access to clean water gives communities more time to grow food, earn an income, and go to school -- all of which fight poverty.

EDUCATION - Clean water helps keep kids in school, especially girls. Less time collecting water means more time in class. Clean water and proper toilets at school means teenage girls don't have to stay home for a week out of every month.

WOMEN EMPOWERMENT - Women are responsible for 72% of the water collected in Sub-Saharan Africa. When a community gets water, women and girls get their lives back. They start businesses, improve their homes, and take charge of their own futures.

We work with local experts and community members to find the best sustainable solution in each place where we work, whether it's a well, a piped system, a BioSand Filter, or a system for harvesting rainwater. And with every water point we fund, our partners coordinate sanitation and hygiene training, and establish a local Water Committee to help keep water flowing for years to come.

Join the team and help us give clean water and so much more to earth's inhabitants. (https://my.charitywater.org).

and water, the document had this to say:

Significant changes in water quantity and quality are evident across the country. These changes, which are expected to persist, present an ongoing risk to coupled human and natural systems and related ecosystem services. Variable precipitation and rising temperature are intensifying droughts, increasing heavy downpours, and reducing snowpack. Reduced snow-to-rain ratios are leading to significant differences between the timing of water supply and demand. Groundwater depletion is exacerbating drought risk. Surface water quality is declining as water temperature increases and more frequent high-intensity rainfall events mobilize pollutants such as sediments and nutrients.[90]

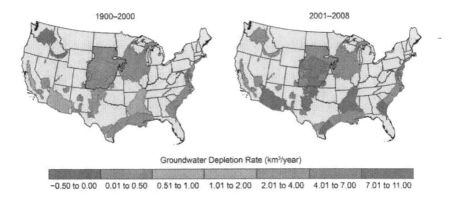

(Left) Groundwater decrease 1900-2000
(Right) Groundwater decrease 2000-2008

(left) Groundwater supplies have been decreasing in the major regional aquifers of the United States over the last century (1900–2000). (right) This decline has accelerated recently (2001–2008) due to persistent droughts in many regions and the lack of adequate surface water storage to meet demands. This decline in groundwater compromises the ability to meet water needs during future droughts and impacts the functioning of groundwater dependent ecosystems (e.g., Kløve et al. 2014). The values shown are net volumetric rates of groundwater depletion (kms3 per year) averaged over each aquifer. Subareas of an

---

[90] FOURTH NATIONAL CLIMATE ASSESSMENT: CHAPTER 3: WATER
https://nca2018.globalchange.gov/chapter/3/

aquifer may deplete at faster rates or may be actually recovering. Hatching in the figure represents where the High Plains Aquifer overlies the deep, confined Dakota Aquifer.[91]

You may be unmoved and arguing the validity of all this and be ready to poo-poo the facts away for not having a direct impact on you. Hang on to your water-sac babe. According to World Vision USA, globally, 844 million people lack access to clean water. Without clean, easily accessible water, families and communities are locked in poverty for generations. Children drop out of school and parents struggle to make a living. Women and children are worst affected — children because they are more vulnerable to diseases of dirty water, and women and girls because they often bear the burden of carrying water for their families for an estimated 200 million hours each day.[92]

In an article by LiveScience, "6 countries (Brazil, Russia, Canada, Indonesia, China, and Colombia) have 50% of the world's freshwater reserves."[93] According to NASA satellite data, 21 of the world's 37 largest aquifers have reached negative sustainability as more water was used than replaced. According to NASA satellite data, 21 of the world's 37 largest aquifers have reached negative sustainability as more water has been used than replaced. Considering 35% of our potable water comes from aquifers, our species is writing checks the planet can't cash. In places like China, India, the United States and France, the future of potable water is questionable at best.[94]

In early 2018, Cape Town, South Africa prepared its residents for, "Day Zero," an event to occur in April which would render Cape Town the first major city in the world to run out of water. Through massive feats of

---

[91] Source: adapted from Konikow 2015.4 Reprinted from Groundwater, with permission of the National Groundwater Association. © 2015.
[92] World Vision: Global water crisis: Facts, FAQs, and how to help
https://www.worldvision.org/clean-water-news-stories/global-water-crisis-facts
[93] Live Science: How Much Water Is on Earth? https://www.livescience.com/29673-how-much-water-on-earth.html
[94] AGU100: Quantifying renewable groundwater stress with GRACE
https://agupubs.onlinelibrary.wiley.com/doi/full/10.1002/2015WR017349

conservation and good humaning, Cape Town was able to stave off the crisis, presumably until "Day Zero," in the winter of 2019[95]. They are rushing to complete water treatment plants and desalination plants, but that brings us to the extinction crisis. We'll discuss the catastrophic disruption of biodiversity taking place on Earth in a bit but you need to know that the process of desalination can destroy billions of microscopic lives that other species depend on for survival. Observer.com reported, in the same article, that Cape Town is not the only major city on the freshwater shit-list. The list includes Tokyo, London, Cairo, Sao Paulo, Beijing, and Mexico City. Still not worried?

It will be Texans, Floridians, Georgians, Californians, and residents of Utah who will face a dry future without immediate action.[96]

Around the world, the effects of climate change are already taking their toll on average citizens. In the summer of 2019, Chennai, India's main reservoir ran dry and more than 4.6 million people had lost their main sources of drinking water. I can imagine some super dystopian future where the only way to get water is to buy it on the black market, but sadly for many citizens of all classes in Delhi, that hellish reality is now.

The BBC reports that around 20% of Delhi's population does not have access to piped water and rely on water tankers for delivery to a central location. Many poor are unable to transport containers larger than a few gallons and supplies are usually depleted within 15 minutes of arrival. The difference between demand and supply is more than 207 million gallons a day which forces millions to rely on the black market - water supplied by private contractors. This network of illegal providers has come to be known as "the water mafia," and they dig illegal wells or

---

[95] Times Live: Fear did more than water restrictions to defeat Day Zero
https://www.timeslive.co.za/news/south-africa/2018-12-05-fear-did-more-than-water-restrictions-to-defeat-day-zero/
[96] The Observer: These American Cities Are Running Out of Water
https://observer.com/2018/02/los-angeles-atlanta-miami-and-san-francisco-are-running-out-of-water/

siphon water directly from leaky pipes.[97]

Anuj Porwal, a social activist from Sangam Vihar, says,

It's always existed here; a network of people who steal water and then supply it to those who need it and they have the backing of the police and local politicians. That's why no one can stop them.

These providers charge between 20-50 rupees ($0.3-$0.7) per bucket while around 60% of the Indian population earns less than $3.2 per day.

In the Middle-East, circumstances are nearing the critical stage. Eight of the 10 most water stressed countries in the world are in the Middle-East. In 2018, reports by the United Nations Development Programme (UNDP) and the Pakistan Council of Research in Water Resources (PCRWR) issued a warning that Pakistan will reach absolute water scarcity by 2025.

In Palestine, residents suffer such water shortages that they rely on water tanks stored on rooftops. The tanks are mostly black and the Palestinians call this "the Black Forest."

Being a California girl, I remember a drought when I was in the 5th grade. We weren't allowed to water our yards. The acres of almond orchards behind my house were given barely enough water to stay alive, and at school, we were only allowed to visit the water fountain as we left the playground. I remember that clearly because my all-time favorite teacher, Mrs. Kimoto appointed me conservation monitor. My first gig as a bouncer was as playground water fountain conservation enforcer.

In 2014, hundreds of wells near Bakersfield, California ran dry. Until late 2016, local residents were forced to go without running water. Trucks delivered bottled water until a nearby town extended its water supply. The World Resources Institute reports that the U.S has 10 states with high or extreme water stress (New Mexico is the imperiled, followed by

---

[97] BBC NEWS: Water mafia: Why Delhi is buying water on the black market
https://www.bbc.com/news/world-asia-india-33671836

California). No matter how you feel about climate change, its causes and implications...it will eventually make its way to you.

If I based my assessment of planet Earth from space, I would think it was a water planet. Yes, 70% of the surface is water, but less than 3% is fresh water; remembering that less than 1% of that amount is accessible to humans. Almost 70% of that water is in ice caps and glaciers, and around 30% is groundwater. Only a tiny bit, 0.3% is in lakes, rivers and swamps. That makes 99% of Earth's water unusable to sustain human life.

It's easy to see rivers and lakes receding, but aquifers are like life banks of water underground, where it's difficult for most of us to imagine them running dry. I find the whole mechanism of the aquifer a dreamy and magnificently efficient design. Our planet was designed with life in mind. Every system is like the tooth of a gear; it all works in perfect synchronicity to keep you alive and Tweeting. But over population and corporate greed are depleting even ancient aquifers faster than they can recover.

In a report by the U.S. Government Accountability Office (GAO) entitled, FRESHWATER: Supply Concerns Continue and Uncertainties Complicate Planning, it states:[98]

In essence, we know that 40 of our 50 states expect shortages, but the degree of a shortage is presently unpredictable.

Do you see that Montana is the only entire state expecting a statewide water crisis? It's no coincidence that Montana exceeds the national average for usage in irrigation and livestock, (keeping in mind that most of the crops grown are to feed cattle.) A staggering 96.3% of the state's water is used for agriculture.[99]

---

[98] U.S. Government Accountability Office (GAO)
FRESHWATER: Supply Concerns Continue, and Uncertainties Complicate Planning
GAO-14-430: Published: May 20, 2014. Publicly Released: May 22, 2014
[99] USGS Science for a Changing World: Water Use in Montana
https://www.usgs.gov/centers/wy-mt-water/science/water-use-montana

Extent of State Shortages Likely over the Next Decade under Average Water Conditions, 2013

Shortage category	Number of states in each category 2013
Statewide	1
Regional	24
Local	15
None	8
No response or uncertain	2

Sources: GAO analysis of state water managers' responses to GAO survey; Map Resources (map).

State Shortages Likely Over the Next Decade

Now we get to nonpoint source pollution. As we learned in the definitions, this is the contamination of waters when the natural cycles of hydrologic movement, (snow, rain etc.) pick up human-made pollutants and taint lakes, rivers, wetlands, coastal waters and groundwaters.[100] Try not to freak out about what I'm going to tell you but there is no way to farm enough land to feed all the humans we will have in 25 years with the diet we presently consume. Agriculture is literally killing the planet on every environmental front. Deep breath…we have lots to share with you in *Solutions*, so the fat lady ain't singin' yet. Here's what agriculture has to do with nonpoint source pollution:

Agriculture accounts for 70% of total water consumption worldwide and is the single-largest contributor of non-point-source pollution to surface water and groundwater. Agriculture intensification is often accompanied by increased soil erosion, salinity and sediment loads in water and by the excessive use (or misuse) of agricultural inputs (e.g. fertilizers) to increase productivity. Pollution caused by agriculture can contaminate water, food, fodder, farms, the natural environment and the atmosphere. Pesticides and fertilizers used in agriculture can contaminate both

---

[100] EPA: What is nonpoint source pollution?
https://19january2017snapshot.epa.gov/nps/what-nonpoint-source_.html

groundwater and surface water, as can organic livestock wastes, antibiotics, silage effluents, and processing wastes from plantation crops. Pollution caused by large-scale industrial farming (including livestock and fisheries) is categorized as point-source pollution, and pollution caused by small-scale family-sized farming is considered non-point-source pollution.[101]

I'll tell you this again; a 1.5 acre piece of fertile land can produce 37,000 pounds of plant-based foods or 375 pounds of meat. Right now, plant-based crops are predominantly being grown to be fed to livestock. My darlings, until cellular agriculture is out and affordable, (our best bet and you'll hear the skinny in *Solutions*) please, please, please eat less industrial meat.

Montana sued Wyoming over water, Florida sued Georgia for the same, and in the meantime, populations are booming; populations which are consuming animal products responsible for one-third of the potable water used. It's a vicious and deadly cycle.

Prior to the city of Detroit, Michigan filing bankruptcy, in an effort to curb the hemorrhage of money and resources, the city cut off water to 80,000 of its citizens; many but not all of them were impoverished. 80% of Americans live paycheck to paycheck and are one crisis away from losing everything. In the event of an unexpected expense, often the only way to keep above water is to sparse out a paycheck along several lines of credit. Utility companies don't work that way, but most are willing to take partial payments for a given period until the customer catches up. In Detroit, all accounts past due were called in and shut off without notice. Imagine your life without water for a single day. When you wake up in the morning, what's the first thing you do? The toilet doesn't flush, there's no shower, no water to brush your teeth, no water for coffee and nothing to drink. Now, imagine 273 of those days strung together. The situation was grave enough and inhumane to the degree that the United Nations

---

[101] Water Action Decade: Water pollution from and to agriculture
https://wateractiondecade.org/2017/12/09/water-pollution-from-and-to-agriculture/

said this:

...a failure to do so, (restore water services) would be a violation of the most basic human rights of those residents.

Detroit's water crisis didn't stop there. In a desperate struggle to hang on, the prices for Detroit water skyrocketed. To curb *their* water fund shortfall, the city of Flint, Michigan decided to use local water. When Flint changed its water source from treated Detroit Water and Sewerage Department water (sourced from Lake Huron and the Detroit River) to the Flint River, officials did not apply corrosion inhibitors to the water in an effort to save the city $140 per day. Ultimately, this caused mass contamination of lead in the city's drinking water that leached from lead in the pipes. Officials persistently dismissed claims that Flint's water was making people sick until the residents themselves took action.

Most of us have heard the words "Flint, Michigan" and "water crisis" at some point but you should understand what really happened so you see how easily this could happen to you and your family. I am sharing the crisis summary from the Nation Resource Defense Council:

Long before the recent crisis garnered national headlines, the city of Flint was eminently familiar with water woes. For more than a century, the Flint River, which flows through the heart of town, has served as an unofficial waste disposal site for treated and untreated refuse from the many local industries that have sprouted along its shores, from carriage and car factories to meatpacking plants and lumber and paper mills. The waterway has also received raw sewage from the city's waste treatment plant, agricultural and urban runoff, and toxics from leaching landfills. Not surprisingly, the Flint River is rumored to have caught fire—twice.

As the industries along the river's shores evolved, so too did the city's economy. In the mid-20th century, Flint—the birthplace of General Motors—was the flourishing home to nearly 200,000 people, many employed by the booming automobile industry. But the 1980s put the brakes on that period of prosperity, as rising oil prices and auto imports resulted in shuttered auto plants and laid-off workers, many of whom eventually relocated. The city found itself in a precipitous decline: Flint's population has since plummeted to just 100,000 people, a majority of whom are African-American. About 45% of its residents live below the poverty line. Nearly one in six of the city's homes have been abandoned.

This was the lay of the land in 2011, when Flint, cash-strapped and shouldering a $25 million deficit, fell under state control. Michigan Governor Rick Snyder appointed an emergency manager (basically an unelected official chosen to set local policy) to oversee and cut city costs. This precipitated the tragic decision in 2013 to end the city's five-decade practice of piping treated water for its residents from Detroit in favor of a cheaper alternative: temporarily pumping water from the Flint River until a new water pipeline from Lake Huron was built. Although the river water was highly corrosive, Flint officials failed to treat it, and lead leached out from aging pipes into thousands of homes.[102]

Did we learn from this catastrophe and is your city prepared to ensure your water is safe? Most of us just don't have the energy to worry about it and that's what caught the residence of Newark, New Jersey off guard. In August, 2019, the world found out that lead-contaminated drinking water is threatening the health of Newark's residents. The levels of lead in their drinking water are some of the highest recently recorded by a large water system in the United States. The city has had the greatest number of lead-poisoned children in New Jersey.

Water, as it turns out, is not a basic human right; it's a privilege entitled to those who can afford it. Okay, let's talk about that. What are water-rights in the U.S.? Built into the real estate infrastructure of the U.S., certain entitlements come with the ownership of land and that includes rights to use the various forms of water that exist on a property.

Riparian rights are given to landowners whose property is located along

---

[102] NRDC: Flint Water Crisis - Everything You Need to Know
https://www.nrdc.org/stories/flint-water-crisis-everything-you-need-know

flowing water, i.e. rivers and streams. Landowners have the right to use the water as long as it does not cause harm to the waterway. The landowner owns the land *beneath* the water to the exact center of the waterway. With this entitlement, the landowner may make use of the water within their needs for drinking, livestock, bathing and watering gardens. These are considered domestic uses. It is important to note that water rights are regulated state-by-state and it is up to the state and municipality to decide if the water may be pumped or removed from the source. By the *Rule of Capture*, each landowner may capture as much groundwater as they can put to beneficial use; therefore, well-owners are not obliged to other landowners for taking their water. The intent behind this liberal dispersal of precious resources is to encourage economic development, which leads us to our present situation with corporate giant, Nestlé.

In August 2019, Nestlé applied for a permit to pump more than one million gallons of water a day from the pristine, Ginnie Springs in Florida. Historically named for a woman named Virginia who washed clothes in the springs, it has been a source of life and respite since the Timucuan Indians inhabited this area prior to European expansion. Jacques Cousteau once referred to the crystal clear water as "visibility forever." Since 1971, six springs and the surrounding 600 acres have been owned by the Wray family. While Ms. Wray and her son own and operate the springs as a recreation area, her two daughters owned Seven Springs Water Company. In 1999, the Wray sisters were awarded a permit to extract 1.152 million gallons of water per day, but in the last four years have only reached 0.2659 million gallons a day; that is, until Nestlé stepped in. In January, Nestlé purchased the plant which means the permit is only a renewal, not a new application. This makes fighting the permit more difficult.

To better understand the specific risks associated with extracting that amount of water from the Santa Fe River, I consulted the non-profit organization, Our Santa Fe River. Michael Roth explained in terms even I could understand. Now that Nestlé has entered the picture with CHF

91.43 billion (2018) in their bellies, bulk transfer will enable the plant to use the full 1+ million gallons a day despite the fact that the Santa Fe river is considered in recovery. In the Gainesville Sun, during an interview with Merrillee Malwitz-Jipson of Our Santa Fe River she said:

Our Santa Fe River is very concerned about bottling any drops of water out of the Santa Fe basin. What is their need, and how can they justify extracting that amount of water when it's never been done before?

The article also notes that spring flows have been impacted across the region by increased pumping of the aquifer to handle the booming population, particularly on the Atlantic coast. White Springs, for instance, is largely dry after once being a thriving tourist spot.[103]

And it doesn't stop there. You may be interested to know that the CEO of Nestlé had this to say to you:

The one opinion, which I think is extreme, is represented by the NGOs, who bang on about declaring water a public right. That means that as a human being you should have a right to water. That's an extreme solution. The other view says that water is a foodstuff like any other, and like any other foodstuff it should have a market value.

In tiny Osceola Township, Michigan, population 900, Nestlé filed a permit to build a pumping booster station which would increase the rate of extraction from Twin Creek River presently at 250 gallons per minute to 400. The local residents have already seen the deterioration of the waterway which has increased water temperature and diminishing trout populations.

Claiming that it inherited rights to forest water dating back more than a century, in 2016 alone Nestlé took nearly 32 million gallons of water from wells and water collection tunnels in the forest considered the Strawberry Creek watershed; *this* in a state already plagued by drought.

---

[103] The Gainesville Sun: Permit sought for bottled water from Ginnie Springs
https://www.gainesville.com/news/20190801/permit-sought-for-bottled-water-from-ginnie-springs

Our Santa Fe River, Inc.

FRIENDS OF THE SANTA FE RIVER

Our Santa Fe River, Inc. is a not-for-profit 501(c)(3) organization incorporated in Florida on December 18, 2007. Our all-volunteer organization is composed of concerned citizens working to protect the waters and lands supporting the aquifer, springs and rivers within the watershed of the Santa Fe River in North Central Florida by promoting public awareness pertaining to the ecology, quality, and quantity of the waters and lands immediately adjacent to and supporting the Santa Fe River, including its springs and underlying aquifer.

We do this via three channels:

Advocacy – members of our advocacy team engage in ongoing letter writing campaigns to the government entities in charge of environmental protection to make them aware of the dangers facing our river systems, to remind them of their protective responsibilities and to recommend actions that will beneficial our watershed. We also appear and speak regularly at water management board meetings, county commission meetings, state environmental agency meetings, and meetings of other environmental groups.

Stewardship – our stewardship team forms clean-up parties several times during the year gathering groups of ten to thirty volunteers to walk the banks of segments of the Santa Fe River and its tributaries or kayak the river equipped with grabbers and net bags, cleaning from hundreds to thousands of pounds of trash out of the waters and watershed lands each time.

Education – we table at various fairs and festivals throughout our area with a booth equipped with literature and demonstration pieces to help the public understand the karst construction and recharge mechanisms of our region and underscore the importance of keeping pollutants out of our waters. We speak at civic group meetings with a presentation designed to facilitate understanding of the delicate balance of our natural systems.

Over the years we have attempted to stop or at least mitigate the effects of water bottling plants, agricultural and industrial over-pumping, phosphate and other mineral mining, concentrated animal feeding operations, the spraying of poisonous chemicals to control vegetation and the use of biosolids, among other issues. We make agriculture and industry aware of the natural constraints on their activities in places that must be sustained to provide for the continued existence of us all.

We invite everyone to our website (www.oursantaferiver.org) for more information, and encourage everyone to subscribe to our newsletter.

So, in essence, we Americans have multiple water mafias operating under the loving wing of corporate-entitlement. They purchase land that authorizes them access water which they bottle, (in plastic) and sell for a profit. These are water sources on which our future generations may need to rely on for survival. They are perfectly entitled to remove as much as they like, for free, bottle it, store it and sell it back to us in our time of need. If you pick up a bottle of Nestlé water anywhere in the country, look at the "source" of the water. It almost always comes from some metropolitan water district – i.e., local city water.

In the worst-case scenario, if the region where you live suddenly faced the loss of water utilities, what would you do?

In a really straight forward and illuminating article by Brian Khan of Gizmodo, *A Global Water Emergency Is Right Around the Corner— Unless We Stop It*, he gives the most practical advice you'll find on solving our American water crisis:

Water managers can design reservoir systems and water use plans that conserve enough water in dry years and capture the runoff in wet years. Setting aside money to fix leaky, lead-infested pipes could also offer another major avenue to conserve water. Water overuse is something that can also be overcome by improving crop and soil management practices or citing energy infrastructure in places with less water stress. Some of the solutions will require creativity and political will, but striving for them is better than the thirsty alternative.

I think that's a great place to end this chat on our water situation. In our upcoming chapter *Solutions*, you'll read lots of things you can do in your daily life to conserve water and even some cool technologies for producing and storing your own. Like, did you know that washing your dishes by hand averages around 27 gallons of water while most energy efficient dishwashers only use 3? I don't know about you but please give me a reason to let the dishes wait!

My loves, if we have proven anything to ourselves as a species it's that we are great at adapting and evolving when we need to. Well, darlings, this is the time. Human-up!

Ocean

Did you know the sea produces 50% of the world oxygen? She gives us oxygen, food and holds within her, an entire world teeming with color and life we have yet to discover. Every single life on Earth is directly connected to the sea. What do we give her in return? We slaughter her children; drag nets the size of villages that kill everything in their path and we dump mountains of garbage in her belly every day, starving off the life that is already struggling to survive. Her bones, the coral reefs, are being suffocated and her shores turned black with oil. We're the worst kids ever.

Geeeeez Pamela, lighten up. I know, I know. I don't want to stress you out more than you already are. You have enough on your plate just trying to survive and provide the very basics for your family. But dammit guys, these are our Earthling-siblings out there and they need us more now than ever. You have a direct impact on everything they're going through right now. If that doesn't guilt you into becoming a more aware consumer then let me take a shot at demonstrating how the hellish nightmare our ocean is living is already affecting your life, your future and that of your children…*our* children.

Overfishing

Something I say often is, "Take care of the ten feet of dirt around you and the Earthlings within your reach to protect." Addendum: make that dirt, water, mud…whatever you're standing on, take care of it. I believe we must return to a simpler way of life comparable to indigenous methods in order to survive the darkness ahead. That having been said, in our natural habitat we are predators. Animal protein was the necessary component of our evolution that caused our brains to grow and create complex problem-solving skills. When we take only what we need for our family and village we're doing our job. Personally, I'm all about keeping the playing field as level as possible. In survival situations, kill what you can, any way you can. In a time of plenty, maybe hunt with a bow instead of a long distance, high powered rifle. As natural creatures, the physical effort of the hunt, the processing of the animal and food preparation is what burns calories and keeps us from storing too much fat. But most importantly, this puts you within proximity of your prey to look it in the eye before you kill it. If you don't have the balls to do that, to take responsibility and have respect for the life you are ending, you have no conscionable right to eat meat. What does all that have to do with overfishing?

When we fish to feed our families, we pull one life at a time out of the water and it has the chance to fight us and break free. But industrial trawling is the most gluttonous and irresponsible method of hunting on this planet. A trawl net is a large wide-mouthed fishing net dragged by vessels along the bottom or in the midwater of the sea or a lake.[104] When I picture the bottom of the ocean, I imagine the place where I grew up, the Sierra Nevada Mountains. The Earth rises up into awe inspiring peaks, dips down into lush, green valleys which were hand carved by Mother Earth over millions of years of blissful solitude; a landscape of abundant life above and below the surface. Now imagine two huge planes flying overhead dragging a 330 foot by 40-foot net, bulldozing everything in its path and scooping up *all* life it touches, as its dragged 100 feet through an unspoiled forest…under the sea. These nets are indiscriminate killers, capturing millions of tons of sea life. Oceana reports that as much as 40% of the catch is bycatch.[105] Bycatch is the non-target fish and ocean wildlife and its one of the biggest threats to our ocean ecosystems globally. As much as 63 billion pounds of innocent lives are discarded, dead or dying, back into the sea. Over 300,000 sharks, sea turtles, dolphins and porpoises, birds, and even seals are

---

[104] Lexico: Definition of Trawl http://english.oxforddictionaries.com/trawl
[105] Bycatch Report Final: WASTED CATCH: UNSOLVED PROBLEMS IN U.S. FISHERIES https://oceana.org/sites/default/files/reports/Bycatch_Report_FINAL.pdf

killed every year by this atrocious practice.[106] The Maritime Executive reports:

Industrial fisheries that rely on bottom trawling to catch fish threw 437 million tons of fish and $560 billion overboard over the past 65 years.

Weighted nets plow along the seafloor destroying already vulnerable coral reefs, mangroves and kelp forests, living creatures (some thousands of years old that serve as nurseries for baby fish), and destabilizing the entire ecosystem. Trawling and deep-sea mining physically disturb the sediment and disrupt carbon sequestration, which can re-suspend stored carbon into the water. They ravage anemones, sponges, sea pens, urchins, and other fine, fragile-bodied animals. For the trillions of shelled or soft-bodied animals like worms, amphipods, clams, crabs, lobsters and many others that live on the seafloor, this means a slow crushing death. For the sea creatures able to escape this horrific death, the food on which they rely to survive, has not. Scientific American tells us:

OSLO, Sept 16 (Reuters) - The amount of fish in the oceans has halved since 1970, in a plunge to the "brink of collapse" caused by over-fishing and other threats, the WWF conservation group said on Wednesday.

Populations of some industrial fish stocks, such as a group including tuna, mackerel and bonito, had fallen by almost 75%, according to a study by the WWF and the Zoological Society of London (ZSL).

We are literally killing the sea. If you eat fish from a package, you're contributing. I'm hearing Sylvia Earle again, "No blue, no green…" game over team humans. Please support Oceana in their campaign to establish science-based sustainable fishing:

https://oceana.org/our-campaigns/responsible_fishing/campaign

Setting and enforcing science-based limits to govern how much fish is allowed to be taken out of the seas has been shown time and again to

---

[106] World Wildlife Foundation: Bycatch victims
https://wwf.panda.org/our_work/oceans/problems/bycatch222/bycatch_victims/

allow fish populations to remain healthy and, in many cases, dramatically increase in size. Oceana seeks to win policy victories around the world that put in place and enforce science-based catch limits.

Overfishing is exacerbated by harmful fishing subsidies. These payments cause too many boats to be on the water and encourage fishing beyond sensible reason. Oceana works in Europe and elsewhere to limit these subsidies.

In addition, many industrial fishing vessels also operate unlawfully in the worlds' oceans, engaging in IUU fishing. Without knowing how much fish fishing vessels catch, and which types of species they are landing, scientists cannot create scientifically-based fishing quotas that can allow species to be fished at responsible levels while continuing to grow the size of their populations.

Climate Change

The ocean is our buffer between us and rampant climate change. Not only does it have everything to do with life as we know it, but the ocean is a key element of our planet's thermostat. I'm a huge fan of the National Resource Defense Council (NRDC) as they are one of the go-to places for stats and numbers on all things environment, (stats…purrrr go my geeky loins.) A very cool cat by the name of Sam Wicks at the NRDC has answered some pretty bizarre questions for me in my quest to understand our impact on the planet. In May 2019, Lauren Kubiak wrote a terrifying but incredibly relevant article based on the findings of *The Global Assessment Report on Biodiversity and Ecosystem Services*, written by an intergovernmental body of biological and ecological experts representing 50 countries and it paints a grim picture of the dire situation for species richness across the globe. The report finds that between half a million and one million species are threatened with extinction globally. The primary mover? Yep, humans. In it we learn that besides overfishing, another reason marine biodiversity is in decline is climate change.

Fish biomass is expected to decrease up to 25% by the end of the century. If you live in a coastal community, this will dramatically affect your life. If you're alive right now, it will directly affect you too. If you're not alive, how about some help keeping my dog in the yard. Seriously. Are you very busy? I digress. Climate change…yes even the dead will find this disturbing. Since the 1970's, the oceans have absorbed 93% of excess heat generated by climate change and is a natural reservoir that stores carbon-containing chemical compounds (a carbon sink) for nearly 30% of human-caused CO_2 emissions. The increase in temperature is driving fish towards the poles but rapid sea ice loss and acidification are likely to prevent biodiversity increases in polar waters. And in the tropics, many local species are expected to go extinct. Coral reefs are projected to undergo extreme warming events with less recovery time in between, causing large bleaching episodes with high levels of mortality. Heavy sigh.

If all that's not bad enough news, climate change is warming the ocean which causes it to expand (thermal expansion) Land ice and glaciers are melting so fast that when coupled with thermal expansion many islands and coastal cities are literally sinking under rising seas. Due to rising sea levels and the over-extraction of groundwater, Indonesia has said the country would be relocating its capital city, in part because it's sinking into the Java Sea. Jakarta is one of the fastest sinking cities in the world. And Houston, Texas has been sinking for decades. Like Jakarta, the over-extraction of groundwater is partly to blame.

The sinking of land resulting from groundwater extraction is called *Sugsidence* and it is a growing problem in the U.S. and around the world. Without adequate pumping regulation and enforcement groundwater is being depleted faster than it can regenerate. One estimate has 80% of serious U.S. land subsidence problems associated with the excessive extraction of groundwater. The Houston Chronicle reported that parts of Harris County, which contains Houston, have sunk between 10 and 12 feet (about 3 meters) since the 1920s, according to data from the U.S. Geological Survey. Areas have continued to fall as much as 2 inches per year.[107] We'll talk more about our melting glaciers just ahead in *Cryosphere*, which is portions of Earth's surface where water is in solid form, including sea ice, lake ice, river ice, snow cover, glaciers, ice caps, ice sheets and frozen ground including permafrost.[108]

Every aspect of our existence is touched by the ocean. Did I say that already? Let me say it again: the ocean is you. It's not too late, my loves, but we need to implement strong protections in at least 30% of our ocean and employ sustainable fishing practices right now in order to prevent more drastic biodiversity loss. Look forward to more remedies in *Solutions* but if you're ready to take action, support the NRDC in their efforts to stop the proliferation of climate change:

https://www.nrdc.org/issues/climate-change

NRDC is tackling the climate crisis at its source: pollution from fossil fuels. We work to reduce our dependence on these dirty sources by expanding clean energy across cities, states, and nations. We win court cases that allow the federal government to limit carbon pollution from cars and power plants. We help implement practical clean energy solutions. And we fight oil and gas projects that would pump out even more pollution.

More Dead Zones

No, not the gaps in cell service that makes me so angry I spew chains of profanity until I run out of them and shout, "Incumbent bikes!" We're talking here about ocean hypoxic dead zones which are an increasingly widespread condition in the ocean (and fresh water) with low or depleted oxygen. We already discussed nonpoint source water pollution, well now you get to see what that turns into. NOAA has an awesome podcast

---

[107] CNN Travel: Indonesia's capital city isn't the only one sinking
https://www.cnn.com/2019/08/27/world/sinking-cities-indonesia-trnd/index.html
[108] Wikipedia: Cryosphere https://en.wikipedia.org/wiki/Cryosphere

HOPE SPOTS: CRITICAL TO THE OCEAN'S HEALTH

I wish you would use all means at your disposal — films, the web, expeditions, new submarines, a campaign! — to ignite public support for a network of global marine protected areas, hope spots large enough to save and restore the ocean, the blue heart of the planet. – Dr. Sylvia Earle's 2009 TED Prize wish that launched Mission Blue

Today, Mission Blue inspires action to explore and protect the ocean. Led by legendary oceanographer Dr. Sylvia Earle, Mission Blue is uniting a global coalition to inspire an upwelling of public awareness, access and support for a worldwide network of marine protected areas – Hope Spots. Under Dr. Earle's leadership, the Mission Blue team implements communications campaigns that elevate Hope Spots to the world stage through documentaries, social media, traditional media and innovative tools like Google Earth.

Mission Blue also embarks on regular oceanic expeditions that shed light on these vital ecosystems and build support for their protection. Currently, the Mission Blue alliance includes more than 200 respected ocean conservation groups and like-minded organizations, from large multinational companies to individual scientific teams doing important research. Additionally, Mission Blue supports the work of conservation NGOs that share the mission of building public support for ocean protection. With the concerted effort and passion of people and organizations around the world, Hope Spots can become a reality and form a global network of marine protected areas large enough to restore the ocean, the blue heart of the planet.

Hope Spots are special places that are critical to the health of the ocean — Earth's blue heart. Hope Spots are about recognizing, empowering and supporting individuals and communities around the world in their efforts to protect the ocean. Dr. Sylvia Earle introduced the concept in her 2009 TED talk and since then the idea has inspired millions across the planet. While about 12% of the land around the world is now under some form of protection (as national parks etc.), less than six% of the ocean is protected in any way. Hope Spots allow us to plan for the future and look beyond current marine protected areas (MPAs), which are like national parks on land where exploitative uses like fishing and deep sea mining are restricted.

Without the blue there can be no green (mission-blue.org).

series, *NOAA Ocean Podcast*, that you can find on their website and one of the episodes was *Dealing with Dead Zones: Hypoxia in the Ocean*:

When water runs off of farmland and urban centers and flows into our streams and rivers, it is often chock-full of fertilizers and other nutrients. These massive loads of nutrients eventually end up in our coastal ocean, fueling a chain of events that can lead to hypoxic "dead zones" — areas along the sea floor where oxygen is so low it can no longer sustain marine life.

And we're back to burgers. I thought I would never hear the end of "carb loading" in the 90's. Now, not only are carbs against the moral code of beautiful people, but in the ocean we have to worry about *nutrient loading*. Nutrient loading/pollution is the process where too many nutrients, mainly nitrogen and phosphorus, are added to bodies of water and can act like fertilizer to nourish the soil. We have so overused and abused our land than now we have to put nutrients back into the soil to grow food with any nutritional value. Farmers apply nutrients on their fields in the form of chemical fertilizers and animal manure, which provide crops with the components necessary to grow and produce the food we eat. But the excess of what is not absorbed returns to our waterways via…pop quiz: A type of pollution we talked about in water. Anyone? Yes, nonpoint source pollution. All those nutrients are washed from farm fields and into waterways during rain events, and when snow melts it can also leach through the soil and make its way into groundwater. High levels of nitrogen and phosphorus in our water brings us full circle back to hypoxia ("dead zones").

Hypoxia kills bottom fauna and the poor shellfish and worms can't move fast enough to escape so they die too. This is a disruption of the food chain supporting the entire eco-system. Every little thing is a link in the biodiversity chain. The fish that do manage to survive a dead zone by escaping or that have had brief exposure escaping a predator, well they suffer for it. Many of them have reproductive and growth impairment and some even change sex. *So, I was walkin' tru da Holland Tunnel and* when I come out da udda side, BADA BING, boobs an a innie. Honest doc, I was a regular guy dis mornin'. Yeah, that has to suck.

It also affects the local economy. Brown shrimp is a deliciously big deal on the Gulf Coast and it has taken a major economic hit from the dead zone there. In 2017, the dead zone was the size of New Jersey.

Nutrient loading, also occurs by wastewater discharge from sewage treatment plants. All of that together feeds algae which stimulates growth, forming an *algae bloom*; a rapid growth of algae in freshwater or marine water systems. The organisms die and degrade and the bacterium that eats them consumes mass amounts of oxygen…yanno, the stuff we breathe that the ocean produces 50% of our supply. Now don't let this give algae a bad rap. Algae are a vital component of our environment and may even save us as a food source. But too much of anything is seldom a good thing. (Except comedy. I hope I die laughing and preferably not while shouting *watch this!*)

In that informative NOAA podcast, Alan Lewitus, director of the Competitive Research Program for the National Centers for Coastal Ocean Science, explains that climate change is exacerbating dead zones mainly due to three factors: oxygen is less soluble with higher temperatures so less of it dissolves into the ocean; marine life consumes more oxygen because higher temperatures contribute to higher metabolic rates; and higher temperatures lead to more stratification, meaning the more oxygenated surface water doesn't mix well with more hypoxic bottom waters. It's a vicious cycle.

It's all connected. All of life on this planet is connected. Every choice we make and every mistake we allow to be made on our watch; it's all our responsibility to change and to exact change where needed **today**.

Pollution

Plastic is the devil. Okay, yeah I touch it every day knowing what it's doing to our planet, but try to live one day without touching a single piece of plastic. Presently, it's almost impossible. Even some Amish communities have begun to use it. *No! Don't do it! Turn back!* We learn from Conservation International (Conservation.org) that 8 million metric tons is how much plastic we dump into the oceans each year. That's about 17.6 billion pounds or the equivalent of nearly 57,000 blue whales. By 2050, ocean plastic will outweigh all of the ocean's fish. Plastic is wreaking such havoc as a pollutant that it has its own place in this book in a chapter ahead. So let's move on to other types of ocean pollution.

The hands-down worst polluter on our planet is that nonpoint source pollution as it accounts for 80% of pollution to the marine environment. That means the worst of it comes directly from you and me. Besides the damage done by agricultural runoff, every human on the planet contributes. Each squirt of insect repellant, sunscreen; every drop of oil, coolant and fluid to drip on the roads and driveways, and even the insecticide you sprayed on a hornet's nest; every chemical we use ends up somewhere and the runoff from our roads, rivers and drainpipes finds its way to the sea. Even the clothes you wear have a massive impact. Each time you wash your clothes, more than 700,000 synthetic microfibers are whooshed into our waterways. Unlike natural materials such as cotton or wool, these plasticized fibers don't break down. One study showed that synthetic microfibers make up as much as 85% of all beach trash. You see my loves; we each hold the power in our collective numbers to make monumental change and save our species from a horrific demise. For just about every chemical we use in our daily lives, there is a natural alternative. We'll show you some innovative and affordable products and gadgets in…you got it, *Solutions*.

There are other forms of pollution most of us would not even consider had we not encountered it first hand or had someone point it out. Allow me.

Thermal Pollution

Thermal pollution is the harmful release of heated liquid into a body of water or heat released into the air as a waste product of a business. An example of thermal pollution is water used for cooling in a power plant that runs into a nearby river and harms the river's ecosystem.[109] Power plants and industrial/manufacturing plants extract water from local lakes and streams, use it to cool machinery, and then they return it with all that heat it absorbed.

Thermal pollution can also be caused by **soil erosion,** which is the degradation of the upper layer of soil caused by natural things such as the movement of water, ice, snow, air, plants and animals but humans can cause it via deforestation, excavation and agriculture. Sediment is deposited in lakes and rivers, making the water murky, and that makes it difficult for light to penetrate the water. This causes problems for aquatic plants that need sunlight for photosynthesis. Sediments are also rich in nutrients such as phosphorus and nitrogen and we know how bad those are in over-abundance. But that sediment also elevates the water, causing it to heat up. (Shallow water heats up faster.)

Even the water that runs off of pavement returns to streams, lakes and rivers not only full of pollutants, but warmer. We touched a little on what warmer water does to the eco-systems while chatting about climate change and hypoxia, but let's take a closer look. Much in the way climate change has affected the ocean, there are many of the same issues that thermal pollution causes:

- Reduced dissolved oxygen (DO) levels
- Loss of biodiversity
- Mass deaths
- Mass migration
- Release of toxins such as radiation

Because warm water is unable to retain as much oxygen as cold water, the rapid increase in temperature literally suffocates marine wildlife and plants. When dissolved oxygen concentration is low in the water

---

[109] Your Dictionary: Thermal Pollution
https://www.yourdictionary.com/thermal-pollution

(anoxic), sediments release phosphate into the water column and algae find this stuff delicious. As we already learned about these guys, algae blooms die and the bacterium that eats them consume more oxygen.

Biodiversity is a delicate tree of lives…a circle, if you will (thank you Simba). When one link in the food chain dies, that condemns their food or predators to their fate. Warmer temperatures are lethal to many fragile ecosystems and thermal shock kills plants, insects, amphibians and fish, alike. When one domino goes down, the cascade of lives will follow.

Those that survive book-it, they move out in a mass migration to the next, better hunting grounds and living space. We humans know how that turns out: precious resources dwindle down rapidly and what ensues is a fight for survival.

In the United States, about 75 to 82% of thermal pollution is generated by power plants.[110] The remainder is from industrial sources such as petroleum refineries, pulp and paper mills, chemical plants, steel mills and smelters.[111] In a 2017 study by Jacelyn Rice, Duke University, Durham, and Paul Westerhoff, Arizona State University, Tempe, the authors determined that wastewater discharges make up more than half the water flowing in many U.S. streams.[112] While all wastewater does not create thermal pollution, most carry contaminants.

(The authors) combined data on more than 14,000 water-treatment plants with hydrology data on the streams the treated wastewater is discharged into. They found that in more than 900 of these streams, the discharges made up over 50% of stream flow. In a subset of 1,049 waterways for which detailed flow data were available, the authors found that when water levels dropped, 635 streams exceeded safe

---

[110] Laws, Edward A. (2000). *Aquatic Pollution: An Introductory Text*. New York: John Wiley and Sons
[111] EPA, Washington, D.C. (May 2014). "Technical Development Document for the Final Section 316(b) Existing Facilities Rule."
[112] Nature Geoscience: High levels of endocrine pollutants in U.S. streams during low flow due to insufficient wastewater dilution https://www.nature.com/articles/ngeo2984

concentrations of at least one endocrine-disrupting compound.

In the face of a wave of deregulation for U.S. big business, wastewaters may become more dangerous than they were in the 1970's. Sadly, President Trump's EPA is feverishly dismantling the Clean Water Act. The proposed changes even by EPA's conservative estimates suggest that more than half of the streams in the country could lose protection.

The Southern Environmental Law Center is fighting to protect U.S. waters from these detrimental changes and you can support them on social media via #ProtectCleanWater. To take action, visit:

https://www.southernenvironment.org/protect-southern-water

Acoustic Pollution

One thing about we humans on which all our Earthling sibs would agree is we're noisy feckers. In the ocean, this is particularly a problem because a large percentage of sea life depends on sound for survival. In my quest to understand, I found the coolest website called Discovery of Sound in the Sea (Dosits.org) run by University of Rhode Island, Graduate School of Oceanography. I had no idea that many species of fish make sounds and some rely on sounds, like the sound of a coral reef, to navigate and survive. There's evidence that underwater reef sounds may be detected by coral reef fish and invertebrate larvae, guiding them to coastal areas and allowing them to identify acceptable settlement habitats.[113] Fish and sounds...that reminds me of a funny story: I was visiting my daughter in Tampa and her home was on a lake. After being up for 48 hours writing, my wolf-dog Tlahda and I took a walk down to the dock for a break. As I got to the edge of the dock, I stepped on a fish in the grass that screamed, waddled on its fins down the dock and gave me shit-eye as it jumped in the water. That was the exact sentence I said to my daughter, a vet tech at the time. There was a sweet smile, "Hmmf.

---

[113] Simpson, S., Meekan, M., McCauley, R. and Jeffs, A. (2004) Attraction of Settlement-Stage Coral Reef Fishes to Reef <u>Noise</u>. *Marine <u>Ecology</u> Progress Series*, **276**, 263–268. https://doi.org/10.3354/meps276263

Cool mom," and a very elegant dismissal. Later that night she came running in my room with a video playing on her phone, "Mom, is this what you saw?" I'll be damned if there isn't an amphibious creature called a mudskipper that does just that, screech and all. I was redeemed.

On Dosits.org, I also learned that animals in the sea rely on sounds to convey a great deal of information quickly over long distances. Variations in rate, pitch, and the structure of sounds can communicate different messages. Fishes and marine mammals use sound to communicate messages associated with reproduction and territoriality. Some marine mammals also use sound to keep the family or pod together. Marine mammals can use echoes (sound bouncing off a target) to detect objects underwater; a process called echolocation. They can use it to find food and even determine the size and shape of an object, its location, whether it is moving, and its distance. Not only can they use it to locate prey, it also helps whales and dolphins analyze their environment. Now this…this is the greatest visual: A cleaner shrimp announces itself as a cleaner and advertises its services by clapping one pair of its claws when reef fish approach. *Chica chica chica boom, I clean you!* What? I'm seeing Carmen Miranda as a shrimp. These guys are awesome as they help the fish on the reef stay clean by removing dead skin and parasites from their bodies.

When was the last time you experienced complete and total radio silence for any length of time? I mean no cell, no Wi-Fi, no TV or radio; complete silence. I'm watching the horrors of hurricane Dorian's aftermath in the Bahamas right now and among the horrific images was that of a man stranded who only wanted his sister in the U.S. to know he's alive. It's been several days, but must have seemed an eternity for them both. Well that's what human activities are doing to our marine life siblings in the ocean.

An organization with which we volunteer as a team and as individuals is the Center for Biological Diversity. For several years now, they have been a consistent and reliable source of identifying environmental problems and their solutions. On their site, BiologicalDiversity.org regarding *Ocean Noise*, they explain that the loudest and most disruptive sounds we humans make in the ocean come from military sonar, oil exploration and industrial shipping.

Naval sonar systems work like acoustic floodlights, sending sound waves through ocean waters for tens or even hundreds of miles to disclose large objects in their path. But this activity entails deafening sound: Even one low-frequency active sonar loudspeaker can be as loud as a twin-engine fighter jet at takeoff.

The seismic exploration of offshore oil and gas drilling, and the ongoing noise generated by the commercial shipping industry essentially cuts of the sea life communication network and disables many of their evolutionary survival mechanisms.

Please visit

https://www.biologicaldiversity.org/campaigns/ocean_noise/

and lend your support for their ongoing campaign to protect our marine life from these lethal threats.

We're determined to save imperiled whales and other marine mammals from acoustic disturbances and sonar-caused mortality — what prominent biologist Sylvia Earle has called "a death of a thousand cuts."

Acidification

Since I never made it to senior year chemistry, I learned about the importance of neutralizing acidity in pasta gravy (spaghetti sauce) from my mother in-law during our 6am Sunday morning cooking sessions, by sprinkling just a touch of sugar to mellow the acidity and balance the complexity of a good tomato sauce. Considering the number of Italian dishes structured around the sauce, this was a life and death skill. Right now in the ocean, our gravy is ruint!

The figure to express the acidity or alkalinity of a solution is considered its pH, the lower the number, the higher the acidity. Before the industrial age, when we loaded up the atmosphere with all that CO_2, the average ocean pH was 8.2 - now it's 8.1. Big deal, right? Actually, it is. Each unit decrease of one pH is a ten-fold increase in acidity. That means the ocean pH is 26% higher now than in preindustrial times.

Besides CO_2, other acid-forming compounds are released into the atmosphere when fossil fuels are burned and this contributes to coastal acidification and those pesky algal blooms. That chemical stew created by burning fossil fuels is released into the atmosphere and falls back to the Earth's surface in the form of acid rain. Acid rain has a pH between 4.2 and 4.4, so when this rain falls in coastal waters, the waters are excessively injured by acidity.

All that CO_2 the ocean has absorbed trying to keep our biosphere in balance formed carbonic acid. This acid reacts with calcium carbonate which is an essential mineral component of seashells. As seawater becomes more acidic, there is less carbonate available for animals to build shells and skeletons. In areas of high acidification, shells and skeletons can begin to dissolve.

I'll leave you with that mental image.

Coral Reefs

Coral reefs have been called the rainforests of the sea, and scientists estimate that as much as 25% of all marine species live on and around them. In terms of biodiversity, this is where the cool kids hang out. Our reefs act as nurseries, protecting babies and affording them a chance to grow. These remarkable invertebrates have a symbiotic relationship with a type of algae called zooxanthellae which live in the corals and provide their tough exterior that is made of the calcium carbonate we talked about above. Zooxanthellae also provide corals with their fancy colors.

As the temperature of the water increases, zooxanthellae are expelled from the coral, causing it to lose its color and a major source of nutrition. This is called coral bleaching, and it's a huge problem for our oceans. Our reefs are being decimated by these temperature increases as well as acidification in that they are unable to build their exoskeletons in acidic environments.

The warming seas and increased acidity may begin to dissolve our reefs if the present atmospheric CO_2 concentrations of 390 ppm reach the projected levels of 560ppm by 2050. But there are things you can do right now to help stop the nightmare. Eat less meat, eat sustainable fish, use less transportation, choose anything but plastic when shopping and please support one of my very favorite organizations, Force Blue Team. They are actively engaged in missions to protect and restore the world's coral reefs.

The founders are Rudy Reyes, a gorgeous, real-life aqua man, and my friend Jim Ritterhoff. Jim is one of the most genuine and talented humans you'll ever meet. He's an avid diver, lifelong environmentalist and a helluva guy. Rudy is a veteran of Iraq and Afghanistan, U.S. Marine, 1st Recon Battalion. He's the original seed for the term BAMF (look it up). These guys worked together and created an organization to serve our beloved veteran Special Operations Forces by giving them a mission to save our planet through conserving and restoring our seas. *Giving warriors a Cause and a Cause its Warriors.*

Please support, follow, engage and LOVE **ForceBlueTeam.org**

Cryosphere

When scientists refer to the cryosphere, they're talking about all the parts of natural Earth that are frozen. Although fossils and core sampling can tell us a lot about what happened during the last great thaw, the dinosaurs didn't leave us any contemporary or primary evidence to prepare us for ours: how rude of them. Our post-industrial CO2 dump on the planet has never been done before in the way that we've done it so we're figuring out this mess as we go along; and while the worst offenders continue to dump it. In 2015, the journal Nature published a study by a team of researchers in Australia and the United Kingdom that gave us a pretty clear idea that about every 100,000 years we experience freezing and thawing scenarios. But how and why are not the big questions here. Our biggest concerns are that we have accelerated the present thaw by altering our atmosphere. The World Climate Research Programme did a great job of highlighting the most disconcerting global consequences of our melting cryosphere:

- Thawing permafrost and the potential for enhanced natural emissions of CO_2 and methane to the atmosphere formerly trapped in the permafrost;
- Shrinking of mountain glaciers and large ice sheets with consequent sea-level rise and impacts on water resources; and
- Declining coverage of sea ice and snow, which will affect marine and ground transportation across the Arctic.[114]

Permafrost is ground that remains frozen for two or more consecutive years. It's made up of rock, soil, sediments, and ice which holds the elements together. Some permafrost has been frozen for tens or hundreds of thousands of years. Permafrost is covered by a layer of soil and can be from 3feet to 4,900 feet thick. Here's the critical point for you to know: It stores carbon-based remains of plants and animals that became frozen before they had a chance to decompose. Scientists estimate that the world's permafrost holds 1,500 billion tons of carbon, almost double the amount of carbon that is currently in the atmosphere.

Unfortunately, when permafrost warms and thaws, it releases carbon dioxide and methane into the atmosphere. As the global thermostat rises, permafrost, rather than storing carbon, could become a significant source of planet-heating emissions.[115]

---

[114] World Climate Research Programme: Melting Ice and Global Consequences
https://www.wcrp-climate.org/grand-challenges/gc-melting-ice-global-consequences
[115] Stated of the Planet – Earth Institute: Why Thawing Permafrost Matters
https://blogs.ei.columbia.edu/2018/01/11/thawing-permafrost-matters/

That brings us to those melting glaciers and land-based ice-sheets that, combined with thermal expansion, cause sea levels to rise. In May 2019, the NRDC reported on a study that reveals:

Global sea level could rise by as much as six and a half feet by 2100, swamping coastal communities and displacing nearly 200 million people worldwide, unless we roll back the tide of climate change.

More than half the surface area of the Arctic is ocean and most Arctic creatures rely on the ice-based eco-system. For instance, polar bears give birth on the ice. They spend their lives traversing the vast spans looking for food and a little sumthin' sumthin' (wink wink nudge nudge). Mothers must load up on the calories in the spring to carry them longer distances in winter to find food. This is the whole key to their survival. Getting stuck in one place because the sea ice has melted means inevitable starvation for the whole family.

Some seals such as the spotted seal, harp seal and ringed seal rarely come to land, relying wholly on Arctic sea ice to give birth and raise pups.

Many seabirds rely on sea ice for fishing, and scavenging and in Canada, the ivory gull numbers have declined by 90% in the past 20 years.

Sea ice edges are where walrus hunt. They use the ice like diving platforms and dive down to the sea floor to feed on clams. Floating ice is also their preferred mode of long-distance travel to better feeding spots.

Even that good algae we discussed forms the base of a unique web of marine life. The warming temperatures and salinity of surface waters is disrupting the whole eco-system down there.

And humans…many communities have relied on marine life for survival for thousands of years. The indigenous seem to be suffering the worst of our consequences when, ironically, they are the only humans who have successfully lived in harmony with our natural world.[116]

Cities and coastal regions are literally sinking all over the world. Our beautiful islanders are being washed away like ants off a rock. But, the purpose of this book is *not* to frighten you. The entire reason that Paul and I have spent the previous year of our lives writing this for you is to tell you, it's not too late. I know this is scary. I know you want to say *fuck it* and just live a life of hedonism until it's over. But my darling Earthlings please, please hang on.

For the last two days, I am recalling a conversation with my youngest child in which I learned she believes we've passed the tipping point. While I am eyeballs deep in all this research, I'm hearing my very brilliant, very educated child with degrees in international affairs and political science with a minor in history explain in a very calm tone, the end of existence will occur in her lifetime. I've stopped to cry every hour since.

I'm going to tell you exactly what I told my own beloved child because this book is her birthday present, my life's work and an offering to my magnum opus; the collective of my four children. I'm writing 10 hours a day right now to be sure I get it right. As I am researching and processing this information, I am repeatedly bursting into tears to recall the vivid images my child painted of how she sees her place in this dying world. But you, my beloved child and my beloved fellow Earthlings, you have the potential to literally change the world. My heart is utterly broken to think you have consigned yourself to doom.

But I will not give up. I won't stop. I won't believe that this beautiful planet is not your future. You are the essence of how good we are and how much we deserve to continue. Don't give up. Stand up, put up your fists and *fight* for the right to exist. The corporations creating this planetary crisis can only sell what we continue to buy. Don't buy it! Begin to imagine the power of our numbers and rather than taking to the

---

[116] Green Facts: Arctic Climate Change https://www.greenfacts.org/en/arctic-climate-change/l-2/5-arctic-animals.htm

streets in rebellion, where they can pick us off like rabbits, use your quiet super human power and make different choices in your quiet and beautiful life. It's your life. Take it back. It's your planet, protect her.

Chapter 4: Biosphere

The place between heaven and Earth where life can be found is the biosphere. On planet Earth, just above and below her surface, life abounds (including humans). Technically, we live in our own space called the anthroposphere but right now, 83% of the terrestrial biosphere is under direct human influence. There are few lifeforms on which we humans have not had an impact, and very few of our influences have been beneficial to any species of life or the resources on which they depend.

200,000 years ago, there were less than half a million humans emerging in the biosphere. Today, there are nearly 8 billion. I've always thought breeding should require training and licensing before issuing a permit to create a human. Most of us have no idea what we're getting into until the first kid is around 13. That's when reality kicks in. It requires more training to run an ice cream shop than it does to contribute to the human population. Honestly, I'm grateful that my millennial children have chosen not to breed. I have to be the only grandmother on the planet elated with only two (perfect) grandchildren (and a grandpiggy, grandpups and grandkitties; also perfect.) So what do 8 billion humans popping up overnight mean to the rest of the biosphere?

In 2002, The Guardian published an article than went hardly noticed by most of us:

Earth 'will expire by 2050'

Our planet is running out of room and resources. Modern man has plundered so much, a damning report claims this week, that outer space will have to be colonized.

The article cites a study by the World Wildlife Fund (WWF) declaring that, at our (then) rate of consumption, the planet would be fully depleted of resources by 2050. Today, you can hardly open a news app without seeing this same foreboding headline. *Shit gettin' real y'all.* It's so overwhelming that *Climate Anxiety* and *Climate Depression* are real things now. I wish I could say those of us who tried to warn you then are happy now you've noticed but it's a problem for us too. I *can* tell you, for me personally, my driving force is to focus on the solutions. *Keep your eyes on the horizon, don't engage the darkness.* I'd rather go out fighting than crying- pretty much sums up my life. But, let's skip over all the depressing details about the innocent lives we are taking with us, (Paul covers that thoroughly in *Species Extinction)* and focus on what you and I are doing to contribute to the problems in our daily lives. Soon, you'll see the *Solutions.*

Your Home

The soil is alive with thousands of organisms. In order to build a new home, hundreds of thousands of living beings, including trees, are killed, and many of them would have fed the lush lawn most homeowners look forward to. When a building comes in, life goes out and must be reintroduced by hand. Many of the plants homeowners choose for their beauty are not native to the environment. This adds the complication of potentially invasive species of plants from the wrong environment, which will require twice as much water to keep alive and may strangle the native environment.

In the U.S., construction and real estate provide 10.3% of the Gross Domestic Product (GDP).[117,118] In order to build the homes and businesses in which we spend our lives, the construction industry uses half of our non-renewable resources. As many as 300,000 houses are demolished annually, which generates 169.1 million tons of construction

---

[117] Statista: Value added of the construction industry as a share of gross domestic product in the U.S. from 2007 to 2018 https://www.statista.com/statistics/192049/value-added-by-us-construction-as-a-percentage-of-gdp-since-2007/
[118] The Balance: Real Estate's Impact on the U.S. Economy https://www.thebalance.com/how-does-real-estate-affect-the-u-s-economy-3306018

and demolition (C&D) debris, which is nearly 22% of the U.S. solid waste load. In the U.K., a quarter of all waste comes from construction. In 2014, 584million tons of C&D debris was generated in the U.S.[119] In 2016, world cement production generated around 2.2 billion tonnes of CO_2, and that's the equivalent of 8% of the global total.[120] Also, as much as 15% of the delivered material is not even used in a project. And what about that noise pollution? Have you ever lived near a construction site? Beep, beep, beep, rumble, BOOM! It's maddening. In addition, globally, construction accounts for 40% of world energy usage. Just like your car, imagine all the components included in the building and construction process; they all had to be manufactured and transported to the site. And what about the natural raw materials such as limestone and water, that had to be excavated in the process? It all adds up quickly. It's one of the least sustainable industries in the world.

But…that's right, there's a *Solution* for that.

Your Car

Before your car ever made it to that fancy showroom, it had already created a massive carbon footprint. Every component, including steel, rubber, glass, plastics, and paints was manufactured and transported, as was the vehicle itself. Unless you are upgrading to an electric car, better to upgrade the one you have and keep her happy. Sounds like good dating advice too.

What about the infrastructure of roads, you ask? Killing animals with vehicles is so common, we have a term for it; *roadkill*. While I once partook in the delicious deer we accidentally killed in the mountains of Utah, there are less expensive and more eco-friendly ways to hunt them. That deer was vindicated by killing our truck. Most animals that attempt

---

[119] EPA: Facts and Figures about Materials, Waste and Recycling
https://www.epa.gov/facts-and-figures-about-materials-waste-and-recycling/construction-and-demolition-material-specific
[120] BBC NEWS: Climate change: The massive CO2 emitter you may not know about
https://www.bbc.com/news/science-environment-46455844

to cross a road don't survive. In fact, road mortality is the leading cause of death and leading cause of species decline for some wildlife populations and an estimated 1 million vertebrates die on roads every day in the U.S Even the animals that got hip to the dangers of roads, are still adversely affected by them. Some snakes will turn around and not cross the road when they encounter it and some animals avoid the surface of the road even when there are no cars on it.[121] This disturbs not only migratory paths, but natural mating patterns. If your Tinder hit is on the other side of a torrential river with no bridge, to what lengths would you go for a *Bit-o-Honey*? Not to mention, okay yeah, you know I'm gonna mention it; the destruction of natural habitats is a huge issue where the species extinction crisis is concerned and 4 million miles of roadway affects 20% of our land. That's a lot of homes destroyed.[122] When you see a proposition come up to build a wildlife tunnel or overpass in your community, please step up and support it. If you don't see such a needed movement locally, make it happen, baby!

Your Groceries

Typical Grocery Store Dairy Aisle

---

[121] Environmental Science: he Environmental Impact of Roads
https://www.environmentalscience.org/roads
[122] Forman, R.T., *Estimate of the area affected ecologically by the road system in the United States.* Conservation biology, 2000

Let me see, plastic, plastic, plastic...ugh, nothing good to report from this dairy isle. Let's drive a few miles down the road, in your electric car, to the farmer's market.

Typical Farmer's Market Bulk Bins

Ah, here we go: fruits, nuts, veggies, spices and even some local meat and dairy. And if you need shampoo, detergent or other household products, you can use CommonGoodAndCo.com to find a refill station near you where you can take your own containers and have them refilled. I love those giant glass pickle jars because we can refill in bulk, saving money and time.

I thought I would start with the good news this time. We both needed a respite from the bad news. My beautiful Earthlings, we all have circumstances that complicate our lives, making time a precious commodity. But your health, the health of your family and the health of the planet are dependent on the choices we make in our *daily* lives, more than those we make at the ballot box. (Although those are important too but let's focus on us first.) Most of us agree that the money at the top, controlled by big business gets the first say in what happens around here. Big business dictates our every move...or do they? Really, it's the other way around. They can only sell what we buy. So, after you get all those delicious foods and household needs from the farmer's market, stop by your big chain grocery store and leave them a letter detailing all the items you had to purchase elsewhere because of their unfriendly packaging standards. The big chains listen to the numbers, as do the manufacturers. When they are paying to store items we no longer buy, they will start selling what we *do* want. It's the infallible cycle of supply and demand. Demand better for your family and planet. I'll go into the details of what you eat in *The Virus Human*.

BETTER MEAT, BETTER WORLD

At Memphis Meats, our mission is to bring delicious and healthy meat to your table by harvesting it from cells instead of animals. You can enjoy the meat you love today and feel good about how it's made because we strive to make it better for you…and for the world.

Cells are building blocks of all food we consume and at Memphis Meats they are the foundation of our approach. We make food by sourcing high-quality cells from animals and cultivating them into meat — think of a farm at a tiny scale. We cut some steps from the current process (like raising and processing animals) and bring nutritious, tasty meat to your table.

By producing meat from the cell level up, we can ensure the highest level of quality at every stage. We aim to keep the benefits of conventional meat while making our products healthier, more nutritious and safer. We want you to enjoy the best of both worlds. It is like having your steak and eating it, too.

We're making meat that is better for animals and that at scale uses significantly less land, water, energy and food inputs. Our process will produce less waste and dramatically fewer greenhouse gas emissions. We believe that the planet will be the ultimate beneficiary of our product.

Please visit our website at (www.MemphisMeats.com) to learn more

Did I mess up by giving you a solution before I told you the problem? I guess I just assume you have realized by now that manufacturing processes and plastic are a huge part of what's killing this planet and our possible future lives. Don't get me wrong, I think it's really clever the way they divided the rows of chocolate chip cookies so all you have to do is pull back the Mylar cover, pour milk, top with whipped cream and binge. (Bad Pamela! Bad, bad packaging!) Okay so, *Hydrosphere* was a stressful and depressing package-of-cookies and box-of-wine chapter. Ugh, plastic. We're coming to that; for now, shop local, shop farmer's markets and refill stations and buy anything but plastic.

Trees

Where I live is quite rural, by choice. I live in a modest home set back off the road surrounded by trees. This is deliberate. I need trees. They are my happy place. Actually, all life on planet Earth needs them. If you remember, I began this conversation on the top of a ridge overlooking a sliver of natural land spared by development. On the day the excavation began, our peaceful home was bombarded with the dull throbbing roar of machines, followed by THUDS that literally shook the house. A few days later, after the sounds had stopped, my wolf-dog Antiope and I took a stroll over to the thick woods across the street to investigate. As we passed the row of tree cover between the 100-acre property and the road, I realized it was being cleared for housing. The first felled tree we came upon was a giant. It lay on its side like a slaughtered elephant. I couldn't contain the wave of terror then extreme sadness. I wailed. I sobbed. This is a creature older than any human present. I walked the length of this beautiful giant, probably 80 feet when she stood upright. It was a horror to behold. She hadn't been cut she had been torn in half by a bulldozer. It was apparent by the multiple gouges that it took several violent impacts to break her and topple her down. I placed my hand on her and bawled with shame for sharing a species with the rumbling monster "progress" that ended her life. I know you think I am anthropomorphizing this "plant," but I probably know more than you about the life of a tree, so please allow me to share this knowledge with you.

- They are parents; they feed and nurture their young.
- They choose the wellbeing of their offspring over their own.
- They deprive themselves nutrients to help a neighboring plant in need, across species.
- They communicate with one another.
- They feel and share pain. If a tree is cut, it releases a signal to the other trees and this "reaction" can be measured in all the trees up to a mile away.
- They are the oldest living land creatures on Earth.

It is because I know these things that I've been having nightmares since the killing began in the woods across the street. I dreamed I was trying to pull a small child from the mud where he was stranded and I couldn't get him out. Then I looked up to see there were dozens of small children stuck in the mud next to the bodies of their dead parents and death was coming for them. There was nothing I could do but cry and fruitlessly tug at their little bodies. But they were going to die.

Another night, I dreamed we had a massive storm and when I stepped out of my house to survey the damage, all my beloved trees were snapped in half and scattered around the grounds like dead soldiers on a battlefield. Maybe I am more tree than human. In the course of my daily research, I am ashamed to be human when I see what we do to other lifeforms on this planet. Why can't we live as the natives do; in peace with the natural world around us that keeps us alive? As determined as I am to leave this planet knowing more than I did the day I came here, the answer to this question I fear will remain a mystery.

North Carolina State University, specifically the College of Agriculture and Life Science, Department of Horticulture Science, has a wonderful page called *Trees of Strength* that is chock full of vital and fascinating facts about trees. I thought I was fairly well-informed on the topic but these guys are a treasure of practical and relevant information about the benefits trees provide. I'm dropping their whole bullet list for you and it's a doozy. Grab a cup-a-joe and check this out:

- Trees can reduce air temperature by blocking sunlight. Further cooling occurs when water evaporates from the leaf surface. The conversion of water to air vapor --- a chemical process --- removes heat energy from the air.
- A tree can be a natural air conditioner. The evaporation from a single tree can produce the cooling effect of 10 room size air conditioners operating 20 hours a day.
- You can improve the efficiency of your heat pump by shading it with a tree.
- Deciduous trees block sunlight in the summer but allow sunlight to reach and warm your home in the winter --- place deciduous trees on the south and west sides of your home.
- Trees can shade hard surface areas such as driveways, patios, building and sidewalks thus minimizing landscape heat load --- a buildup of heat during the day that is radiated at night resulting in warmer temperatures. Ideally, 50% of the total paved surface should be shaded.
- Evergreen trees can be used to reduce wind speed and thus loss of heat from your home in the winter by as much as 10 to 50%.
- Trees absorb and block noise and reduce glare. A well-placed tree can reduce noise by as much as 40%.
- Fallen tree leaves can reduce soil temperature and soil moisture loss. Decaying leaves promote soil microorganism and provide nutrients for tree growth.
- Trees help settle out and trap dust, pollen and smoke from the air. The dust level in the air can be as much as 75% lower on the sheltered side of the tree compared to the windward side.
- Trees create an ecosystem to provide habitat and food for birds and other animals.
- Trees absorb CO_2 and potentially harmful gasses, such as sulfur dioxide, carbon monoxide, from the air and release oxygen.

    - One large tree can supply a day's supply of oxygen for four people.
    - A healthy tree can store 13 pounds of carbon each year ----for an acre of trees that equals to 2.6 tons of CO_2.
    - Each gallon of gasoline burned produces almost 20 pounds of CO_2.

- For every 10,000 miles you drive, it takes 7 trees to remove the amount of CO_2 produce if your car gets 40 miles per gallon (mpg); it will take 10 trees at 30 mpg; 15 trees at 20 mpg; 20 trees at 15 mpg; and 25 trees at 12 mpg)

- Trees help reduce surface water runoff from storms, thus decreasing soil erosion and the accumulation of sediments in streams. They increase groundwater recharge and reduce the number of potentially harmful chemicals transported to our streams.
- An acre of trees absorbs enough CO_2 in a year to equal the amount produced when you drive a car 26,000 miles.
- Trees cool the air, land and water with shade and moisture, thus reducing the heat-island effect of our urban communities. The temperature in urban areas is often 9 degrees warmer than in areas with heavy tree cover.
- Trees can help offset the buildup of CO_2 in the air and reduce the "greenhouse effect".
- Trees create microclimates suitable for growing shade loving plants.
- The American Forestry Association estimates that 100 million new trees would absorb 18 million tons of CO_2 and cut U.S. air conditioning costs by $4 billion annually.
- Dews and frosts are lessened under trees because less radiant heat is lost at night.

Personal and Social Benefits

- Trees are the least expensive plants you can add to your landscape when you consider the impact they create due to their size.
- A tree can add music to your life by attracting birds and other animals.
- A tree can provide pleasant smells. A cherry tree can perfume the air with 200,000 flowers.
- Hospital patients have been shown to recover from surgery more quickly when their hospital room offered a view of trees. They also had fewer complaints, less pain killers and left the hospital sooner.

- Most of us respond to the presence of trees beyond simply observing their beauty. We feel serene, peaceful, restful and tranquil in a grove of trees. We are "at home" there.
- Trees provide us with color, flowers, fruit, interesting shapes and forms to look at.
- Trees can screen unattractive views, soften the sometimes harsh outline of masonry, metal, asphalt, steel, and glass.
- Trees can separate and define space thus providing a sense of privacy, solitude and security, and create a feeling of relaxation and well-being.
- Trees can serve as a living legacy for the next generation, thus linking us to near and distant generations
- Lowered electricity bills are paid by customers when power companies build fewer new facilities to meet peak demands, and they use reduced amounts of fossil fuel in their furnaces and need fewer measures to control air pollution.
- Trees can help direct pedestrian traffic, provide background and thus enhance the appearance of other landscape plants and, our homes.
- Trees help people reflect positively on life's changes.
- Trees have been reported as having a relaxing effect on students studying for exams.
- Studies have documented that urban vegetation can result in slower heartbeats, lower blood pressure and more relaxed brain wave patterns.
- Sound waves are absorbed by tree leaves and branches. A belt of trees 100 feet wide and 45 feet high can reduce highway noise by 50%. Prolonged exposure to noise can cause hypertension, higher cholesterol levels, irritability and aggressive behavior.

Community Benefits

- Trees can create a lasting impression on how a community is perceived by visitors and affect the mood and community pride of its residents.
- Trees can enhance community economic stability by attracting businesses and tourists ---- people linger and shop longer when trees are present.

- Apartments and offices in wooded areas rent quicker and have a higher occupancy rate --- workers in offices in wooded areas report more productivity and less absenteeism.
- Property values of landscaped homes are 5 to 15% higher than those of non-landscaped homes and homes are quicker to resell than homes without trees.
- Trees increase the humidity in the air, help increase groundwater recharge, reduce soil erosion and storm water run-off thus reducing the amount of water we consume and the need for new water treatment plant and storm water structures. A study in Salt Lake City revealed the tree canopy reduced surface runoff by 11.3 million gallons following a 1-inch rain event.
- The feeling of community pride created by trees can help reduce crime.
- By reducing heating and cooling cost, trees can reduce our dependence on oil and natural gas.
- By absorbing and deflecting falling rain, trees can reduce the severity of floods.
- By reducing CO_2, dust and other potentially harmful gasses in the air, our air quality is improved through lower levels of ozone, carbon monoxide and sulfur dioxide.
- Trees make communities livable for people and their activities. People walk and jog more on streets with trees; children and adults have a cool place to play or relax in the summer, thus increasing their interaction with neighbors.

As life forms, how different are we, really? I bet you didn't know that hemoglobin is built around iron (Fe), whereas chlorophyll is built around magnesium (Mg). Our lifeblood is only one tiny rock away from theirs. Below, I am sharing a short-story I wrote for my children many years ago. I updated the story to include the most recent discoveries about the life of trees so for you data freaks, the references to all the studies and white papers are at the end of this story entitled, "Tiny Rocks."

Tiny Rocks[123]

By Pamela Dawn

I feel like reaching far today. Like this will be the farthest I have reached in all my years. This is the day that I will tell my story and you will receive it with warmth and understanding. I am with you now, even as you read this, and soon you will know me as yours; as you are mine.

My name is Jji. The day of my beginning rests to memory now, but I can name my mother and the father of my seed. For what beginnings are not grounded with the profundity of parents? Chatter, really. These first few words are chatter. Beginnings are not relevant unless the middle or the end enliven the senses with curiosity. So, it begins here. I will attempt to touch your senses.

I am 237 years old. I am 83 feet tall and I have lived in the same place my entire life. The same spot…really. I am not a great tree. I have a simple trunk and narrow leaves. But I reach high into the possibility of my growing and reach deep into the womb from whence I sprang.

I have seen you walkers come and go and sometimes find complacency in your passing. I do not wish to offend. But even the walkers I have loved seem strange to me. I do not understand you. Have I intrigued you yet? I want to intrigue you. I think we should get to know each other better. There are things about your life and practices which worry me and I wish to get them off my trunk.

---

[123] Originally published 2/18/2005, iD magazine
https://e360.yale.edu/features/exploring_how_and_why_trees_talk_to_each_other
http://humanorigins.si.edu/research/climate-and-human-evolution/climate-effects-human-evolution
http://www.mycolog.com/fungus.htm
https://www.livescience.com/53618-fungus.html
https://www.sciencedaily.com/releases/2006/10/061021115712.htm
http://science.psu.edu/news-and-events/2001-news/Hedges8-2001.htm
darwin.bio.uci.edu/sustain/bio65/lec02/b65lec02.htm
https://en.wikipedia.org/wiki/Earliest_known_life_forms

Why are you so angry? When in your growing did you learn to feed your mind fat, with mendacious lies? I am sorry for your pain. I take it away when you come near enough. But collectively you bear too much and even I have trouble keeping up. Your pain is like a virus, which you foster, breed and spread. Pain to fear to ego, then to angry ways and no longer can you feel the love I send.

It's not your fault. Equipped with all the higher mammalian attributes, the one you need the most is not inborn. It is learned. Our calling is to teach you. But you are so caught up in your language, you seldom hear our song. I wish to tell you secrets. I wish to show you visions of other worlds. I wish to give you peace and heal your soul. But standing here amidst your deafening campaigns, I can only wait. I have waited long enough. I reach inside your mind, now.

Perhaps because the cord to womb is never severed for us, we remember life as it began. We lived as one back then; one collective life and genetic makeup. Our common ancestor was called LUCA. In the beginning were the Elders-three: Bacteria, Archaea and Eukaryote. Though we all share the structure of LUCA via DNA and RNA, you and I descended from the great Eukaryote. You are my kin. My veins run cold with magnesium-rich Chlorophyll and yours with iron-Blood. When magnesium and iron changed their place and blood began to flow within your veins you were set free from the still silence of a life umbilically tied to the Mother. Some of us think you ran amuck with that freedom. Nonetheless, today these tiny rocks within our veins remain our only difference. I am happy to be still, silent and green.

All mothers wish for their offspring to be so special that the world lifts them up and protects them. The most valued in a species endure with love and when the Mother gave birth to us, we indeed changed the world. 700 million years ago, we took deep gulps of the heavy carbon dioxide air and breathed out oxygen. We trapped the deadly carbon in our bodies and helped cover the rocks with lovely silent snow. This is how you grew strong with the solid structure of a tree inside.

For centuries we lived in harmony. There were even walkers who worshipped our kind and revered our connection to the Mother. But, like spoiled children with no supervision on a playground, you walkers went wild with blood-lust. You forgot our kinship. You desecrated the graves of our ancestors. You cut deep into the body of the Mother, ripped the carbon from our graves and burnt it; setting free the poison gasses to injure all life upon Her. Don't you understand? The mother sent us here with lignin in our bodies to trap the poison far from her new children so you could grow bones and stand upright as we trees do.

I bet you didn't know that we have a developed and functional nervous system. You are so arrogant, you walkers, to think your walker thoughts as if you own them. In fact, our nervous system is very similar to that of your own and like your brain, we have a center called a root neck. We are capable of tactile perception and demonstrate this throughout our kingdom. We grow with tendril vines so sensitive to touch they will begin to coil within twenty seconds. As mimosas, we grow with leaves which droop and appear lifeless when the vibration of an approaching herd is felt. We began as one. We are still akin. But your kind was not happy thinking still so you walked away.

So brave and self-assured that you could live a severed life, you endeavored to create a different world. And that you have. You are very noisy creatures. It seems as though you suffer to be silent with yourself. Maybe you're afraid that if you think too hard and know too much, in your stillness, roots will begin to grow again. Most of the time, I can only imagine what you think. But sometimes, one will come and linger with me; and for a time, your chaos seems clearer. You are frightened children. Deep within, you sense your disconnection and long to be held down by your roots with the certainty of love and simple lives. But that is our purpose. We came to prepare the Earth for you. You are the favored child now, given more freedom and beauty than any life before you. Alas, you are still children with much to learn.

I remember Seane. She was a walker I loved. She lingered with me all the days of her life. For hours she would climb across my branches.

Then, climbing turned to sitting still beneath me. She read to me from books, and then she wrote them. She came to me with vivid dreams and ideas she put to words and then to pages. I gave her my honest opinions and sometimes she would write them with her words. She lived a quiet life below the hillside. She loved and lived unlike most walkers do. Carefree with her gait and every attribute, she was the one I almost envied. The closest I could come to wish for walking was when she danced beneath my swaying bow.

Seane brought each of her children to me on the very day they arrived. She buried one beside me one wet spring. She asked me to watch over her sweet child. Watching turned to living as his little body fed my roots and he became a part of me. I began to think his little thoughts in a tree-like way. Day after day, she laid her withering body against the rock that marked his grave. She poured out pain in waves I almost could not bear. But I will forever cherish the day that Seane heard me thinking my thoughts in his childlike way. She stopped crying to the rock. She stood up on trembling limbs, wrapped her arms around my trunk and sang a lullaby in rhythm with my sway. She spent a season singing to me each dusk and, for the rest of her life, spoke his name as though she understood. She healed the pain of mother-loss with me that season, and never cried about the child again.

Seane's children buried her beside me and beside the rock that bore her child's name. I think her thoughts in a tree-like way now.

But there were others whose thoughts I never care to think and whose ways have left indelible scars on my body. I heard the screams and cries of anguish one cold, fall morning. Reaching with my mind to the crying, I found a forest of my kin. It was a slaughter; a massacre of life and brutal slaying. Limbs were being torn and body's broken, hacked and sawed. Dying trunks were left to decay as lifeless bodies were chopped to bits and carried off by walkers. For many days the crying and the killing filled my senses. And then there was nothing but the silence of life that had once been.

If you want to understand this in terms of your language, we trees communicate in several ways. Through intricate networks of fungal filigrees, chemically, we send signals down complex fungal pathways to warn our kind of environmental change, search for kin, and even transfer nutrients to offspring or a struggling neighboring plant. Sometimes, we do so by denying nutrients to ourselves, in a self-sacrificing act of community survival…even across species. You walkers have yet to learn this.

We also communicate through what you call w-waves. When a tree is injured, there is a measurable response in our entire kingdom for more than a mile away. We are here. We are present. We exist. Can't you imagine our right to exist in peace? I wish that you would.

The walkers of the old ways were much kinder with our bodies. Our wood became their shelter and their fuel and for this, they were grateful to us. They came to us in silence and asked permission. They waited for an answer, sometimes for days. Whenever one was cut, another would be planted. When limbs were taken, food and water were given to our roots. We had an understanding, as the older walkers listened to our thinking. But our beloved friends passed swiftly as a tidal wave of strangers pushed them from their homes and the land we shared. The invading walkers were unkind and had little reverence for life. Eventually, a better man came to settle on our dirt with tiny children and he learned to love our hallowed dirt. He understood the rhythm of the seed to Earth and his family stayed with us for many generations. Eventually, they gave me my Seane.

After the passing of Seane, her children continued to visit with their children, their lovers, and the celebration dances. They grew old and died and their children continued the traditions of their line. But there was only one since her who spoke my language and felt my stillness. He touched the soil with hardened hands and planted the seeds which fed his family as they grew. The old farmer, in the ways of his wise grandmother, spoke to the Earth and respected Her silence. He would come to my side when the weather was bad and give me his fears. He leaned on my body and gave me his pain as his own began to fail. I gave him shade and reassurance and he always left kind words of gratitude behind for me.

The old man is dying now. His son will take the farm and make a field of stone there. I know it won't be long until the season of my time comes to pass and all that I have seen will pass with me and with my friend the farmer.

The old man is dead now. His son has buried the fields in sheets of blackened tar and all the life below has ended. I hear the thunder of machines and feel the low, dull pulse of my impending doom. Soon I will feel the cutting of my body and the scattering of my senses. All that I once was will pass and all that I know with me.

Before I face my final moment, I leave you this. I am with you now. I am all around you, can you feel me? Listen to the sound of tree-song. Feel the possibility of your growing and reach deep into the womb from whence we sprang. I am not a great tree. I have a simple trunk and narrow leaves. But I am only a tiny rock away from being you. In my passing, think your thoughts in a tree-like way.

ENDANGERED EARTHLINGS, INC.

This tree was my sister CayDee's favorite tree and the inspiration

Chapter 5: Atmosphere

Remember way back when I said I wanted you to see the body of Earth like your own body? We are now headed to her lungs. Nope, not the rain forests, they are more comparable to our **alveoli** (tiny, balloon-shaped air sacs sit at the very end of the respiratory tree responsible for the exchange of oxygen and CO_2). The entire **Atmosphere** is her respiratory system and we're giving her COPD like a bunch of chain smoking fools.

If the Earth is our Mother, air is the blanket in which she has all life swaddled. Without air the Earth's surface would be irradiated with ultraviolet radiation and the difference in temperature between night and day would be hundreds of degrees. Not to mention, every organism that breathes would die. But luckily, the Earth has an app for that called the atmosphere. The atmosphere is made up of layers divided by temperature. These include the troposphere, stratosphere, mesosphere and thermosphere and *way* out there; the exosphere.[124]

We live in the troposphere. This is where we sing in the rain, row, row our boats and suffer the wrath of extreme weather. This layer contains 75% of all of the air in the atmosphere, and almost all the water vapor. But not everything that needs air is alive. There are many chemical changes on which the planet relies that are dependent on our delicious recipe for air, which is:

---

[124] NIWA: Layers of the atmosphere
https://niwa.co.nz/education-and-training/schools/students/layers

Name	Chemical formula	Percentage composition by volume
Nitrogen	N_2	78.1 %
Oxygen	O_2	20.9 %
Argon	Ar	0.9 %
Carbon dioxide	CO_2	0.03 %
Neon	Ne	0.002 %
Helium	He	0.0005 %
Methane	CH_4	0.0002 %
Krypton	Kr	0.0001 %
Hydrogen	H_2	0.00005 %
Xenon	Xe	0.000009 %

Thank you, Muhannad Almemar, Quora

Maintaining the right balance of these invisible givers of life is becoming increasingly more difficult for the troposphere and the root of this is human activity. Ever wonder why the airplanes hang around the 30,000 foot mark? This is the beginning of the stratosphere. It gets a bit chilly up there. Up in the stratosphere is the highest concentration of ozone, a colorless unstable toxic gas, very stinky and possessing powerful oxidizing properties. It's formed from oxygen by electrical discharges or ultraviolet light and it does an important job for us Earthlings as it's the stuff that absorbs harmful UV radiation, preventing significant radiation from entering our life-zone.

Much like my delicious red wine, a little UV radiation is good, (helps the skin to form vitamin D), but a lot is…well its way worse than a week-long hangover. It causes sunburn on human skin which can lead to cancer, causes cataracts in our eyes, and can damage the basic building block of life, DNA; even in plants.

Now let's talk about my Papa's hair. Him was so vewy handsome. His salt and pepper hair was one of his best features and he knew it. Thus, he possessed a magnificent collection of aerosol hairsprays. I learned to appreciate his collection, much to his dismay, when I learned it could tame the massive hair-beast on my head my first year of high school. If you were around in the 70's, you probably remember, that was the first most of us learned about the ozone and what the chlorofluorocarbons

(CFCs) used in aerosol sprays, were doing to destroy it. No Bueno. Needless to say, Papa was not happy on that first shopping trip when he discovered his favorite brand missing from the shelves.

There were other culprits like freons and halons, once used in refrigerators and fire extinguishers. Collectively, these guys have reduced the amount of ozone in the stratosphere, particularly at polar latitudes, leading to the "Antarctic ozone hole". This ozone hole (not literally a hole) occurs because of the special atmospheric and chemical conditions that exist there and nowhere else on the planet. The incredibly low winter temperatures in the Antarctic stratosphere cause polar stratospheric clouds (PSCs) to form. Reactions that occur on PSCs, combined with the isolation of polar stratospheric air, allow chlorine and bromine reactions to produce the ozone hole in the Antarctic springtime. Because CFCs last so long, the ozone layer will likely continue to thin into the future.

The *Montreal Protocol on Substances that Deplete the Ozone Layer* (a protocol to the Vienna Convention for the Protection of the Ozone Layer) is an international treaty designed to protect the ozone layer by phasing out the production of numerous substances that are responsible for ozone depletion. The two ozone treaties have been ratified by 197 parties (196 states and the European Union), making them the first universally ratified treaties in United Nations history.[125]

We've covered what this ozone hole means to humans but what are the other implications? The U.S. EPA has this to say:

Effects on Plants

UVB radiation affects the physiological and developmental processes of plants. Despite mechanisms to reduce or repair these effects and an ability to adapt to increased levels of UVB, plant growth can be directly affected by UVB radiation.

---

[125] Wikipedia: Montreal Protocol
https://en.wikipedia.org/wiki/Montreal_Protocol

Indirect changes caused by UVB (such as changes in plant form, how nutrients are distributed within the plant, timing of developmental phases and secondary metabolism) may be equally or sometimes more important than damaging effects of UVB. These changes can have important implications for plant competitive balance, herbivory, plant diseases, and biogeochemical cycles.

Effects on Marine Ecosystems

Phytoplanktons form the foundation of aquatic food webs. Phytoplankton productivity is limited to the euphotic zone, the upper layer of the water column in which there is sufficient sunlight to support net productivity. Exposure to solar UVB radiation has been shown to affect both orientation and motility in phytoplankton, resulting in reduced survival rates for these organisms. Scientists have demonstrated a direct reduction in phytoplankton production due to ozone depletion-related increases in UVB.

UVB radiation has been found to cause damage to early developmental stages of fish, shrimp, crab, amphibians, and other marine animals. The most severe effects are decreased reproductive capacity and impaired larval development. Small increases in UVB exposure could result in population reductions for small marine organisms with implications for the whole marine food chain.

Effects on Biogeochemical Cycles

Increases in UVB radiation could affect terrestrial and aquatic biogeochemical cycles, thus altering both sources and sinks of greenhouse and chemically important trace gases (e.g., CO_2, carbon monoxide, carbonyl sulfide, ozone, and possibly other gases). These potential changes would contribute to biosphere-atmosphere feedbacks that mitigate or amplify the atmospheric concentrations of these gases.

Effects on Materials

Synthetic polymers, naturally occurring biopolymers, as well as some other materials of commercial interest are adversely affected by UVB radiation. Today's materials are somewhat protected from UVB by special additives. Yet, increases in UVB levels will accelerate their breakdown, limiting the length of time for which they are useful outdoors.[126]

While our advances in science and technology are generally good, we make a few mistakes along the way. We're making a lot right now. I think we can all agree: Air, good. No air, bad. Ridiculous Pamela, of course it's bad to not be able to breathe, but I'm here to tell you it's something we're taking for granted and perhaps soon, another CEO will tell us air is a marketable product, not a right.

The World Health Organization (WHO) has some disturbing data to share with us. They report 4.2million deaths every year as a result of exposure to ambient (outdoor) air pollution. 3.8 million deaths occur every year as a result of household exposure to smoke from dirty cook stoves and fuels. WHO estimates that nearly 7 million people die every year from exposure to man-made fine particles in the air. A staggering 91% of the world's population lives in places where air quality exceeds WHO guideline limits. That means 9 out of 10 Earthlings breathe polluted air. If we're breathing it, so are all the other innocent creatures that have the misfortune of sharing a room with filthy, dirty and gluttonous siblings.

There is not a single cause of man-made air pollution, (unless you blame humans in general.) But there are basically two categories; a primary or a secondary source. A factory emitting sulfur-dioxide is a primary polluter. Secondary pollutants are the by-products of the primary pollutants. Smog is a secondary pollutant caused by the interaction of several primary

---

[126] EPA: Health and Environmental Effects of Ozone Layer Depletion
https://www.epa.gov/ozone-layer-protection/health-and-environmental-effects-ozone-layer-depletion

pollutants.[127]

We humans send these contaminants into the air via fossil fuels in factories, cars, airplanes and coal burning electric plants. If you fly, drive or use electricity from the grid, you're guilty. If you use aerosols, pesticides or even smoke, you're guilty. But what may be difficult to imagine is that your diet is a massive contributor to world pollution. Agriculture creates ammonia and this is one of the most dangerous gasses in the atmosphere as well as, livestock creates massive amounts of methane, a greenhouse gas that is 34 times more detrimental than CO_2 to the atmosphere. And guys, it's the cow burps, and not the farts releasing 90% of their contribution of methane. Only 5 to 10% is from manure and poots. Not to mention, the livestock industry uses literally tons of insecticides, pesticides, and fertilizers and they too emit harmful chemicals into the air. We already discussed the major water pollution this industry creates. When was the last time you ate a burger? Your pound of burger also created more than 30kgs of CO_2 and a couple kg's of methane: ONE burger. That's the equivalent of a 200 mile drive. If you are eating a burger that came from the rain forest, when you include the deforestation cleared for livestock, the number goes up to 335kg of CO_2 for one kg of beef.[128] Dudes and duderesses, if you must eat on the road, please get an Impossible Burger at Burger King.

You may have never considered what your daily life indoors does to contribute to global air pollution, or even how your daily and weekly cleaning habits affect your health. Did you ever paint a room then try to breathe in it? Cleaning products can do the same thing. SPM stands for **suspended particulate matter** and it can be anything from dust created when you drill a hole, to spraying those ants with pesticides, to a by-product of combustion from a small engine. Every chemical with which you interact also interacts with your environment and life in it, no matter

---

[127] Conserve Energy Future: What is Air Pollution? https://www.conserve-energy-future.com/causes-effects-solutions-of-air-pollution.php

[128] Simply. Live. Consciously.: Did you know that the production of one burger emits as much greenhouse gas as a drive of nearly 200 miles? https://www.simply-live-consciously.com/english/food-environment/1-burger-200-car-miles/

how tiny.

Here in the U.S., President Trump boasts:

One of the problems that a lot of people like myself — we have very high levels of intelligence, but we're not necessarily such believers. You look at our air and our water, and it's right now at a record clean.

Considering the previous chapter on water, we know that isn't the case and regarding U.S. air quality? In many cities in the U.S., every man woman and child inhales the equivalent toxins to that of smoking 20 cigarettes a day for 29 years.[129] These toxic fumes and particles come to you courtesy of vehicles, planes and factories.

In his incredibly powerful chapter on mining, Paul explained the basics of particulate matter (PM) and Nitrogen Oxide, but to highlight the importance of your learning these terms, I'll repeat some of his thorough information. In a book called, *Every Breath You Take: A User's Guide to the Atmosphere*, Dr. Mark Broomfield explains that it is predominantly fine particles of PM2.5 (particulate matter that has a diameter of less than 2.5 micrometers, 3% the diameter of a hair) and nitrogen oxide, the stuff of rocket fuels, wreaking the havoc on our lungs. Not to be confused with nitrous oxide, the laughing gas…whippits anyone?

To be more specific, PM2.5 includes pollutants, such as sulfate, nitrates and black carbon, and it is this deadly cocktail which poses the greatest risks to our health. They reduce visibility and cause a myriad of respiratory problems. Because it is so small, it passes right through the nose and throat and causes chronic diseases like asthma, heart attacks, bronchitis and other respiratory problems as it makes way into the circulatory system. These tiny invaders require an electron microscope to be detected.

Having lived in them all our lives, most of us are pretty in-tuned to our

---

[129] Independent: Air pollution in cities 'as bad for you as smoking 20 cigarettes a day', says study https://www.independent.co.uk/news/health/air-pollution-smoking-cigarettes-city-research-health-asthma-copd-a9056566.html

bodies. Suddenly, we're running a few less steps in that daily jog, dipping out of the spinning class sooner or staying just a bit longer in yoga child pose enjoying the expansion of lungs filled with air. Marathon runners in polluted cities have significantly lower times. We tend to blame those things on ourselves. "I didn't get enough sleep," "ate too many carbs over the weekend" or, my juicy rationalization of the day, "Damn I'm getting old." What if the sad truth is that we're just not getting the right mixture oxygen in our bodies due to pollution and airborne contaminants? I know for me, allergy season workouts are always shorter and my muscles reach failure sooner. City-dwelling Americans who are just beginning to feel the effects of air pollution must be in a constant state of what-the-hell is wrong with me.

The respiratory system is particularly sensitive to air pollutants because it is made up of a mucous membrane covering its entire internal surface. This amazing engineering was designed to absorb 400 million liters of air over an average lifetime. This begins the transport of oxygen throughout the bloodstream. But those pollutants we talked about can damage that place in the lungs where the exchange of oxygen and CO_2 is produced. Those pollutants are also distributed throughout the body via that bloodstream pathway. Depending on the pollutant, this can directly affect the cardiovascular system and lead to structural damage. In addition, that $PM_{2.5}$ can also cause premature death, asthma attacks, lung cancer, preterm births, autism and dementia. Most recently, a study by researchers in the U.S. and Denmark identified a correlation between air pollution and mental health problems, including bipolar disorder, depression, schizophrenia and personality disorders.[130]

So, how about that "record clean" air? One of my favorite sources for the skinny on what's happening to our resources in the U.S. is the NRDC. In a really informative article, entitled, *U.S. Air Pollution Is "Completely Outrageous,"* I learned that despite our having a clean air law, called the

---

[130] PLOS Biology: Environmental pollution is associated with increased risk of psychiatric disorders in the U.S. and Denmark
https://journals.plos.org/plosbiology/article?id=10.1371/journal.pbio.3000353

Clean Air Act, according to the World Health Organization, the U.S. still ranks #23 in sickness and deaths due to air pollution. That's worse than Honduras (#26) and Nicaragua (#28.) I agree that is outrageous.

In the chapter on the Geosphere, we learned about the deadly repercussions of methane being released via fracking and drilling. Paul also gave you a clear picture of what's happening to the planet as a result of coal mining. In July, 2018, a man named Andrew Wheeler was appointed acting administrator of the U.S. EPA. What you need to know about Mr. Wheeler is that previously he was an energy lobbyist, and among his biggest clients? Murray Energy Corporation, the largest coal mining company in America. It is also important to point out that Murray's CEO, Robert E. Murray, vigorously fought the Obama administration's attempts to reduce carbon emissions and then contributed $300,000 to Trump's inauguration. Mr. Wheeler is also the Vice President of the Washington Coal Club which hosts 300 of the big dogs of the coal industry.[131] He also lobbied the U.S. Department of the Interior to open portions of the Bears Ears National Monument to uranium mining. This is the guy over the EPA.

Shortly after Trump took office, Murray, an unabashed climate denier, presented Vice President Mike Pence with a ridiculously pro-coal "action plan" that called for doing away with the Clean Power Plan (CPP), withdrawing from the Paris climate agreement, eliminating federal tax credits for renewable energy, and—yes—halving the EPA's workforce.[132]

In June, 2019, the Trump administration replaced an Obama signature climate policy, the Clean Power Plan which would have reduced U.S. power sector emissions by 32% by 2030 - with the Affordable Clean Energy (ACE) rule. The new rule aims to lower power sector emissions

---

[131] ABC News: Who is Andrew Wheeler, the ex-coal lobbyist who will become the acting administrator of the EPA? https://abcnews.go.com/Politics/andrew-wheeler-coal-lobbyist-acting-administrator-epa/story?id=56390331
[132] NRDC: Who Is Andrew Wheeler? (And Why You Should Be Afraid of Him) https://www.nrdc.org/onearth/who-andrew-wheeler-and-why-you-should-be-afraid-him

by only 0.7% and 1.5%.[133]

In August 2019, the Trump administration released a detailed plan to cut back on the regulation of methane emissions. This plan proposed a rule to eliminate federal requirements that oil and gas companies use technologies to detect and fix methane leaks from wells, pipelines and storage facilities. Even major energy companies have opposed initiatives to remove climate change and environmental rules.

I know, you're thinking, well I may live in the city but I'm a health-conscious outdoor buff and I hike in the country on the weekends. That oughta fix 'er. Nope. Bad news: In May 2019, the National Parks Conservation Association (NPCA) released a report which found that 96% of our national parks have "hazardous air quality".[134] Sadly, the report found that the elevated levels of air pollution were a direct threat to sensitive species.[135] Air pollution causes lung damage, harm to immunes systems as well as an increase in inflammation. Those rangers living and working in the parks full-time are getting inundated with this crap while serving the wild and her visitors. You may also want to know that the new EPA sheriff just set a plan in motion to open 1.6 million acres in California, in proximity to some of our most pristine national parks, to freaking fracking.

Air pollution is accelerating climate change while climate change is creating air pollution. The current technology to determine exactly how climate change may affect extreme weather events in the U.S.; ways in which atmospheric circulation in the mid-latitudes is impacted, is still under development. We're not even seeing the whole picture yet using

---

[133] VOX: Trump's EPA just replaced Obama's signature climate policy with a much weaker rule
https://www.vox.com/2019/6/19/18684054/climate-change-clean-power-plan-repeal-affordable-emissions
[134] National Parks Conservation Assn: https://www.npca.org/articles/2166-parks-group-s-report-finds-96-percent-of-national-parks-are-plagued-by-air
[135] Parks Group's Report Finds 96 Percent of National Parks are Plagued by Air Pollution https://www.npca.org/articles/2166-parks-group-s-report-finds-96-percent-of-national-parks-are-plagued-by-air

the current climate models. Guys, bad things are afoot in the atmosphere and no one at the top is minding the store. More than 100 million people in the U.S. live in places where air pollution exceeds air quality standards for health. Unless positive changes are made, and quickly, climate change will worsen already vulnerable areas and our health, and our crops and life in these regions will begin to fail. All hell is breaking loose on the average citizens of the world, but politics and greed continue to dictate exactly how and if we deserve to live.

November 8, 2018, in the early morning haze, on just another Thursday, citizens of Paradise, California awoke to a nightmare the whole world would never forget. The burning embers rained down like hail as this sleepy town was transformed to a war zone. Propane tanks were exploding like bombs and skies darkened with a thick black smoke that moved so fast they had little time to conceive the scope of the disaster. When it was finished with Paradise, 86 people were dead and 90% of the town was left a charred ruin.

In another fire in Butte County, California a month later, what began as an electrical fire transmission by PG&E became known as the "Camp Fire." It destroyed 153,336 acres in two weeks, leveled 18,804 buildings and took the lives of 86 people. It was the world's most expensive natural disaster for insurers with losses around $16.5 billion.[136] This is only the beginning. In a July 2019 study entitled, *Earth's Future*, the authors' report that the size of California's wildfires has increased by 500% due to climate change and that could double by 2085.[137] We're not even going into the debate now looming over who pays for Paradise. Let's get back to what Paradise has to do with air quality.

Higher spring and summer temperatures and early spring thaws cause soil to be drier for longer. This increases the likelihood of drought which

---

[136] Patch: Camp Fire Was The Costliest Natural Disaster In the World in 2018 https://patch.com/california/san-diego/camp-fire-was-costliest-natural-disaster-world-2018

[137] AGU100: Observed Impacts of Anthropogenic Climate Change on Wildfire in California https://agupubs.onlinelibrary.wiley.com/doi/full/10.1029/2019EF001210

creates conditions to make the highly improbable incredibly possible. So much so that a farmer mending a fence struck a steel stake with a hammer to plug a wasp's nest and that one little spark ignited a patch of dry grass. That little fire merged with another fire and became the Mendocino Complex Fire…the largest wildfire in California's history. It burned nearly half a million acres before it was finally extinguished four months later but not before it took the life of a firefighter and injured four others.[138]

Ah, Santa Barbara, California. It's one of those places that, even if you live there, you can't help but ooh and awe around each corner of massive rock cliffs and emerald green rolling hills. I feel incredibly fortunate to have taken my driver's training there while living in Ventura. It was like taking a scenic drive with a good friend who is also a totally laid back, pot smoking driving instructor. I'm giggling. He was awesome. Where were we? Ah yes, the air in Santa Barbara. For six straight years, Santa Barbara, California, was on the American Lung Association's list of the cleanest cities in America for short-term particle pollution, in its annual State of the Air report. But in 2019, the city dropped from one of the country's cleanest to one of the top 25 most polluted, in large part to the wildfires. The culprit of the complicating particulate matter in fire-caused air pollution is manmade buildings and cars whose burning releases particulate pollution into the atmosphere. That smoke is often laced with asbestos, heavy metals, and other hazardous chemicals in the polluted air, many we have yet to even identify.

You know how we talked about the rain forests being the lungs of the planet and then I said, "Nah mate," (not sure why I said it with an Australian accent but it happens) I said, "The rain forests are more like our alveoli and the whole atmosphere is Earth's respiratory system". While I'm writing this chapter, I'm watching a horrific event unfold in that absolutely remarkable region of our planet in the Brazilian

---

[138] USA Today: Firefighter dies battling Mendocino Complex blaze in California
https://www.usatoday.com/story/news/nation/2018/08/14/firefighter-dies-mendocino-complex-fire/984613002/

rainforest, the Amazon. I want to inject my feelings here because I'm damn good and pissed. I have a really long fuse but when it goes, I'm downright belligerent toward blatant stupidity. The world-peter running that part of the world, Jair Messias Bolsonaro (38th President of Brazil), is making a mockery of the severity of this catastrophe. He wasted precious time moving into action by whining that his critics were setting the fires to make him look bad, then…oh man. This. I…I can't. So, then, he refused $20 million in aid from the G7 summit because some dude on social media made a comment about another world leader's wife being prettier that his. Yeah…that's my 4th grade version the story, which has now been spun to say the apology was demanded for comments about Bolsonaro's environmental stance. Regardless, that is the degree to which these issues matter to this world leader when a natural resource on his watch is responsible for producing 20% of the Earth's oxygen is burning! By the time you're reading this, either the fires are out or we're all completely fucked. But I'll tell you what we know right now.

The region was already on the verge of a non-reversible tipping point due to deforestation for livestock and on its way to becoming a dry savanna, which, by the way, is how the thousands of fires now burning were started; to clear land for livestock. As a matter of interest, a fifth of their economy comes from agribusiness, and deforestation went up 278% in 2019.[139] I'm not the only one outraged, and in response to massive criticism at home and abroad, Bolsonaro announced he was banning setting fires to clear land for 60 days. He has also accepted an offer of four planes to fight the fires from the Chilean government and has deployed 44,000 soldiers to seven states to combat the fires. There have even been articles published questioning whether any sovereign nation has the right to destroy its own resources when that destruction has global environmental repercussions. Bolsonaro says, "The fire that burns the brightest is that of our own sovereignty over the Amazon," this after accusing European leaders of neocolonialism. To add insult to injury,

---

[139] DW Made for Minds - Brazil: Amazon deforestation rises rapidly
https://www.dw.com/en/brazil-amazon-deforestation-rises-rapidly/a-49925106

recently revealed documents indicate that Bolsonaro not only intends to use hate speech to isolate minorities of the Amazon, (since it worked so well for Trump), but the PowerPoint slides reveal plans to build a hydroelectric plant and a bridge over the Amazon River, as well as a highway in order to "Fight off international pressure for the implementation of the so-called 'Triple A' project."[140] This is an ambitious plan that seeks to join efforts between eight Amazonian countries to protect the Amazon.

Back to our topic, air; what do these fires have to do with air pollution? First, they cause catastrophic damage to forests that would have removed CO_2 from the air. Second, they inject soot and other aerosols into the atmosphere, with complex effects on warming and cooling. This creates a sort of feedback loop, (the greenhouse effect.) As a result, it is estimated that wildfires contribute 5 to 10% of annual global CO_2 emissions each year. And what about that PM stuff? Fires cause outdoor and indoor air to become polluted with both coarse PM10 and fine PM2.5. What we know about wildfires and health is that people who are exposed to wildfire smoke may develop a number of symptoms, including headaches, eye irritation, and irritation of the linings of their nasal passages, sinuses, throat and bronchi. But what about the long-term effects on infants and children? Most scientists don't even know exactly what is in the smoke that wildfires create. Remember, a good percentage of the combustible materials in the Paradise fire were manmade. Also, as smoke rises, the elements inside the smoke change. We may not understand all the long-term implications of wildfire air pollution for years to come. Hopefully, by then, we will have long-term climate solutions up and running.

Not to be a bad-buzz, maker-worser... (Did I lose smart points in that round?) But we still haven't talked about the global air pollution situation. We've talked a lot about what's happening in the U.S. right

---

[140] Open Democracy: Leaked documents show Brazil's Bolsonaro has grave plans for Amazon rainforest https://www.opendemocracy.net/en/democraciaabierta/leaked-documents-show-brazil-bolsonaro-has-grave-plans-for-amazon-rainforest/

now, but that doesn't end our contribution to air pollution globally. Even though, in the past we Americans made strides in environmental improvements, we are not blameless for pollution in the worst contributing countries in the world. The U.S. is no longer a manufacturing economy. Only 8.5% of jobs in the U.S. come from manufacturing. That means, all our stuff is being made, processed or refined somewhere else and that means we are polluting someone else's country. How many Americans boycott clothing made by slave and child labor? Are not our petroleum, plastic and everyday products worthy of the same scrutiny? North America, home to less than 5% of the global population, accounts for about one-fourth of global economic output.[141] That's a lot of stuff.

In 2017, data by WHO evaluated the concentration of PM2.5 in 92 countries and according to this data, Pakistan had the most polluted urban area at a concentration of 115.7ppm, Qatar averaged 92.4ppm and Afghanistan averaged 86ppm. Keep in mind, up to 12ppm is considered "safe." In May 2018, WHO ranked 14 Indian cities amid the world's 15 most polluted cities, Kanpur being the worst based on PM2.5 levels and by PM10 levels. Thirteen cities in India were among the 20 most-polluted cities of the world. Pretty consistently, the worst air in the world is in India.[142]

I want to paint a clear picture for you, of how each of our choices affects the entire world. Kanpur is a great place to start but to complete the big picture will require a trek all around the globe. Welcome to Kanpur. This city has a history of producing superior quality leather that dates all the way back to 1778 when the East India Company made its incursion. Following India's independence in 1947, there were only 13 leather tanneries, but by 2013 there were more than 400. In a hard-hitting photo essay by Arindam Mukherjee of Al Jazeera, October 2013, *Kanpur: A*

---

[141] USA Today: From Bahrain to Qatar: These are the 25 richest countries in the world https://www.usatoday.com/story/money/2018/11/28/richest-countries-world-2018-top-25/38429481/
[142] World Health Organization: Global Ambient Air Quality Database (update 2018) https://www.who.int/airpollution/data/cities/en/

city being killed by pollution, we see disturbing images of rampant toxic pollution (chromium, cadmium, lead, arsenic, cobalt, copper, iron, zinc and manganese) dumped into the Ganges river; twice as much as the city's water treatment unit can process. Our minds imprinted with 5-year old Karan's twisted body, a victim of the deformities among newborn babies and farmers decrying toxic fields. We see bone-thin buffalos grazing on the banks where poisonous chemicals that were used to process their hides flow into the river, a young girl toiling in a contaminated field, plumes of toxic fumes pumping into the air, and 19-year old Iqbal carrying leather on his head to its next treatment.

If we look at Kanpur's exports, you'll see that common theme throughout this book: Animal livestock is a huge part of our daily lives in many, many ways. Let's just examine Kanpur's top 3 exports[143]:

S.No.	Top Products Exported from Kanpur Icd Port	Jan-Dec-2014	% Export Share	Jan-Dec-2015
1	meat of bovine animals, frozen	$232.80M	28.7276	$158.15M
2	leather further prepared after tanning or crusting, including parchment-dressed leather, of bovine (including buffalo) or equine animals, without hair on, whether or not split, other than leather of heading 4114 - whole hides and skins:	$112.98M	13.9420	$51.94M
3	footwear with outer soles of rubber, plastics, leather or composition leather and uppers of leather - sports footwear :	$104.34M	12.8753	$84.23M

Kanpur's Top Three Exports

---

[143] Infodrive India: Major Export products from Kanpur Icd Port & Customs House
https://www.infodriveindia.com/india-ports-customs/kanpur-icd-exports.aspx

Where did these items go? Funny you should ask, that's just where we are headed. If we follow only the exported leather trail, we'll see this graph[144]:

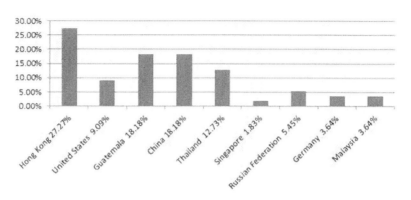

Major Buyer Countries - by Count

Well that's not too bad. The U.S. is only responsible for 9.09% of the leather exports causing the worst of Kanpur's pollution problems. Not so fast my friends. Do you see that Hong Kong takes a whopping 27.27% of that exported leather? It may interest you to know that in this timeframe, the U.S. purchased 35.7% of our finished leather goods from Hong Kong. And China at 18.18%? In 2017, the U.S. imported $7.35billion worth of Chinese leather goods. Are you starting to see the big picture? Chances are, if you purchased anything made with or using bits of leather, you contributed to Kanpur's pollution problem. That awesome record the U.S. has for air and the terrible air quality in India and China? We only looked at one tiny product in the massive influx of manufactured and refined products we use daily that are made elsewhere, and that product created catastrophic pollution that ends up affecting the entire planet. Pollution, in all its forms, does not recognize borders.

Fortunately, as we learn and grow to be better humans, we are pretty great at adapting. Better and stronger regulations in India resulted in only

---

[144] https://www.infodriveindia.com/india-export-data/4107-hscode-export/lp-kanpur_icd-report.aspx

260 tanneries remaining in Kanpur. As a side note, I've had many conversations with farmers about the speed at which our population is growing and the impossibility of continuing to feed humans our present diet. When they say, "My family has been farming this land for generations, what are we supposed to do?" My answer to them is the same I would give to tanners in Kanpur…evolve, my loves. Change has come down upon us hard and heavy and although we didn't ask for it, we helped create these disastrous conditions. We *must* adapt. Now is the time to use everything we've learned to create new and better villages, cities and global communities.

In the U.S. alone, as many as 30,000 deaths were attributed to air pollution in 2015.[145]

As I just demonstrated, every choice we make has a direct effect on us all. We North Americans have deep and lasting impacts on air quality, water pollution and biodiversity all over the world by the power of our consumerism. Every product we purchase, create and sell must have a long chain of sustainable components, or we are condemning the entire planet and all innocence in our wake to disastrous ends. Now is the time for diligence and responsible stewardship of our planet.

---

[145] PLOS Medicine: Particulate matter air pollution and national and county life expectancy loss in the USA: A spatiotemporal analysis
https://journals.plos.org/plosmedicine/article?id=10.1371/journal.pmed.1002856

Chapter 6: The Virus Human

My darlings, there's douchery afoot in the kingdom of Fuckington. Things are bad here on this little rock that was gracious enough to allow us to exist. But I'm afraid she won't tolerate us much longer. We are the Virus Human.

Human Population

So, an egg, a sperm, and a zygote walk into a bar; stop me if you've heard this one. Toss in a little human DNA, Barry White and tequila...we were destined to overpopulate.

While working on a SciFi novel, I learned about an awesome method by which to quantify our development and as we make a study of our habits, it becomes frightening for the future prospects of our species. The DMT (Demographic Transition Model) is a standard by which to measure the rates of birth and death as affected by the region's stage of economic and social development. We're heading into some heavy nerd waters, so put on your horn-rimmed glasses and try to wade through it with me.

Until the industrial revolution in the 1700's, rates of birth and death were both high, keeping the population fairly consistent. This is known as Stage 1. In contemporary populations, this would be areas of the Amazon and Bangladesh. But post-industrial development brings medicine, which lowers the death rates, while birth rates remain high, resulting in a population boom. This is Stage 2, and most under-developed countries are here; such as much of Sub-Saharan Africa, Guatemala, Nauru, Palestine, Yemen and Afghanistan. But as economic changes bring education and improved status for women, birth rates begin to lower. This is Stage 3, the place you will find developing countries such as Botswana, Colombia, India, Jamaica, Kenya, Mexico, South Africa, and the United Arab Emirates, for example.

When a country reaches a place of economic stability, Stage 4, we find strong economies, higher levels of education, better health care, a higher number of working women and fertility drops to an average of two children per woman. This is a Developed country. A few examples are, Argentina, Australia, Canada, China, Brazil, most of Europe, Singapore, South Korea, and the U.S. But something else has emerged in the DMT; humans have begun to produce negative populations in places.

The closest you will find presently to this theoretical Stage 5, is Germany. Were it not for the increased population due to immigrants, the death rate, 11/1,000 out-numbering the birth rate, 8/1,000, would indeed create a negative population. The elderly are outnumbering the contributions of births to the workforce, inciting a whole new problem for the Virus Human.

Why did I drag you kicking and screaming through that delightfully nerdy rant? To say this: When we examine the human population of Earth, the explosive presence of our species is a runaway snowball. The projected population for planet Earth by 2050 is 9.8 billion and no matter how you crunch the numbers, at our present rate of consumption, planet Earth is incapable of supporting 9.8 billion humans. We're down to the final 2 minutes and if *Team Human* doesn't pull a miracle out of their burger-buckets, it's game over.

Food

This is probably the point at which you expect me to come runnin' at you shouting, "Viva la mung bean!" (Not that there's anything wrong with that! Seinfeld? Anyone?) However, I spend 2 - 4 hours a day following the latest information on the topics that most affect our environment and see that our diet has the greatest negative influence. I thought I had a good idea of how the numbers average out across the globe but I'm looking at the tabs open across the top of my browser and holy-monkey is it complicated. The relationship between what we eat and how it affects our planet looks like an obsessive detective's murder pegboard with trails of yarn like a spider web.

The average human on planet Earth eats:

Things We Eat

- 2,870 calories per day
    - 570 (20%) comes from sugar and fat
    - 272 (9%) is meat
    - 235 (8%) is dairy and eggs
    - 327 (11%) is produce
    - 170 (6%) is "other"
    - 1,296 (45%) is grain[146]

But when we look at who is eating what, a single image begins to emerge from the pegboard. Even when you calculate the starving and predominantly vegetarian regions, on average humans consume 6.1 ounces of meat daily.[147] I'm seeing that scene in *My Cousin Vinny* when Marissa Tomei says,

A sweet, innocent, harmless, leaf-eating, doe-eyed little deer. ... Imagine you're a deer. You're prancing along, you get thirsty, you spot a little brook, you put your little deer lips down to the cool clear water... BAM! A fucking bullet rips off part of your head! Your brains are laying on the ground in little bloody pieces! Now I ask ya. Would you give a fuck what kind of pants the son of a bitch who shot you was wearing?[148]

It's bad enough that we eat our fellow Endangered Earthlings, but it turns out where previously it appeared that fossil fuels were the driving factor of global warming, 51% is actually caused by livestock.[149] Let's look at

---

[146] How Much Do Humans Eat By the Numbers? In the United States, It's a Ton https://www.inverse.com/article/38623-pounds-of-food-united-states-calories
[147] National Geographic: What the World Eats https://www.nationalgeographic.com/what-the-world-eats/#
[148] My Cousin Vinny, Starring Joe Pesce and Marissa Tormei
[149] World Watch Institute: Livestock and Climate Change

the other repercussions of our present rate of meat consumption:

- Every second, 1.5 acres of rainforest are cleared for animal agriculture. Even more than for palm oil. (Rainforests are our planet's lungs)
- The meat and dairy industry uses one-third of our planet's limited supply of fresh water
- Animal agriculture is the leading driver of species extinction due to:
  - Deforestation
  - Habitat destruction
  - Ocean dead zones
  - Biodiversity disruption (When one or more species is dependent on another for survival, it creates a cascade effect of extinction when one is lost.)
  - Water pollution
  - Global warming
  - Regional drought

Remember those numbers we talked about, the ones that say you can't sustain 10 billion humans on Earth at the present rate of consumption? Well, check this out: livestock now covers 45% of the available land on Earth. Do you want to talk about the inefficient use of resources? Reminder: A 1.5-acre piece of fertile land can produce only 375 pounds of meat; yet, can produce 37,000 pounds of plant-based food. Presently, if we fed the grain that we feed to cattle to starving humans, we could feed 6 billion people.[150]

If you are an American, on average, you eat 3 burgers per week. When we consider that it took 660 gallons of water to produce that single burger, you waste the equivalent of a

Beer and Cheeseburger

---

http://www.worldwatch.org/node/6294
[150] Cowspiracy: The Facts http://www.cowspiracy.com/infographic

swimming pool full of water (17,280 gallons) every 8 weeks. And that's just the cost in terms of our limited supply of fresh water.

It drives me crazy that people see the term environmentalist next to my name and instantly begin making all the wrong assumptions; tossing around labels like "liberal" and "democrat" and assume I'm a vegan. I am none of those things. I'm an all-American gun toting, flesh-eating Joan Wayne. My father was a casualty of war and my mother was a revolutionary. This entire book is about recognizing the changes we humans have made to the surface of this planet and what we can do about them. I assure you; it will take heroic efforts from every one of us to save the living beings on her. It must begin with the way we eat and grow our food.

Although I preach and teach the environmental perils of eating animals, I am only able to go a couple of years eating solely plant-based foods. There are several reasons, not the least of which is my health. But I am a true reducetarian, eating meat or fish only once a week and from sustainable sources. Thanks to brilliant science and massive funds from the likes of Bill Gates, Richard Branson and Leonardo DiCaprio, real meat is on its way straight from the cow's rump to you via stem cell technology called Cellular Agriculture. Besides vegetable-based alternatives, it's our greatest hope to feed 10 billion humans and everyone I know who has tried it said it's delicious. Grow mama some bacon, guys! As soon as it's out, that's the only meat I'll eat. But I believe we are all headed to extreme life-circumstances that will place average citizens in a fight for their lives. In a survival situation, humans and meat go together like...um, wanting to be alive and actually being alive.

In the beginning, (dun dun dun...nope, not God, still just me), the beginning of humans that is, one thing caused our brains to get bigger, our bodies to get stronger and afforded us the ability to go longer distances to forage: Yep, meat.[151] When a human body is placed in

---

[151] Annual Reviews: Energetics and the Evolution of the Genus HOMO

extreme conditions the need for calories and protein increases. I believe our bodies will evolve beyond the need for other animals as a food source, but I also think, in evolutionary terms, we're not done cookin' yet. Perhaps our great-grandchildren will find the concept of animals for food as alien as we find slavery, but today…it's a choice to be made, not a cross to bear or grounds for condemnation.

Personally, I would love to exist in the naiveté of believing everything is fine, it's all okay, we're gonna be fine. But I have spent the last 12 years obsessed with all things environment and, guys, things are fucked up and getting worse. One of my primary mantras out-and-about is, "take care of the ten feet of dirt around you and the Earthlings within your reach to protect." I consider my optimism about life in general to be an asset, but my rationalism sees what's just around the corner. World Bank has estimated that by 2030, Climate Change alone will send 100 million people into abject poverty.[152] If you are not ultra-rich, then right now you are or know someone who, no matter how hard they work or how much they struggle, they just can't get ahead. No wonder suicide and drug addiction are at an all-time high, and the bad stuff is just starting. In one of my all-time favorite philosophical treatises, *Oh The Places You'll Go,* Dr. Seuss says,

> *And when you're in a Slump,*
> *you're not in for much fun.*
> *Un-slumping yourself*
> *is not easily done*

For some, it's almost impossible. When you get to the point where you've lost a job, car, then home, the chances for finding gainful employment are slight. We are just beginning to see the devastating affects climate change will and *is* having on average humans. As of now, September 2019, we are watching the death toll of people in the

---

https://www.annualreviews.org/doi/10.1146/annurev.anthro.31.040402.085403
[152] Reuters: World Bank warns climate change could add 100 million poor by 2030
https://www.reuters.com/article/us-climatechange-poverty-disaster-idUSKCN0SX0WE20151108

Bahamas at 50 and the missing at 1,300. In his chapter *Climate Change,* Paul will go into depth about how it affects extreme weather and economies. Life is swiftly moving toward a matter of survival.

Growing up in a valley below the captivating but unforgiving Sierra Nevada mountains, we learned to survive first. Even kindergartners learned to curl in a ball and hide under their desk when an Earthquake hit. Then, when we got a little bigger, we learned the true story about cannibalism at Donner's Pass. And instead of summer camp, we went to Summer Survival Camp in Yosemite National Park. I remember then learning that pine needle tea will keep you alive, although I also learned from Euell Gibbons in the old Grape-Nuts commercials that nine parts of the pine tree are edible. Good ole Euell. We miss you, dude. But if you plan to hike off the mountain in extreme weather conditions, it will take more energy than plants can provide. *Yes,* eat vegetarian when you can, but know how to hunt, kill and survive. I guarantee, a week into being lost in the mountains in the winter, you'll crave a burger, not a carrot.

When I married the father of my children, at a very young age, we moved to his family home; a 200-year-old dead farm in Beaver Dam, Utah. He was attending Utah State full time and I was working two jobs, while I simultaneously held the appointed position of POW Representative to Governor Norman Bangerter. We had a baby and the expenses of daycare, so we lived (happily, I might add) in poverty. Were it not for the morning doves and his ability to sell the pelts of the coyotes he shot, we would have been very hungry that first winter on the old farm.

Yes, I am an avid proponent of cruelty-free living and preach the sentience of animals, but as a rationalist, I wish for *you*, my loves, to be prepared. I see our species returning to a more natural way of life, an off-grid harmony and balance with nature sort of life. But we must get there. Stick with me and I'll show you the way.

Please keep in mind; we're still talking about the problems we are facing as a species. This is the place I'm laying out the worst of it, so you'll read the details about Cellular Ag, vegetable-based meats and other problem solving/emerging technologies in our bit about *Solutions*. We'll get there. For now, let's talk about plastic.

Plastic

Thanks to the hard work of dedicated champions of Earth, most humans are aware that we have a problem with plastic. We'll have to treat this problem just like we would any other addiction; one step at a time. The first step is to admit we have a problem. Most of us already know the basics, but just to be sure we're on the same page, I'll spell it out. I know that you know that I know this part sucks…looking at the damage we have caused and the suffering we have inflicted by our addiction; but it will also help to remind us why we need to change. Try to remember that at the end of all the awfulness, there are *Solutions* and healing.

If you have ever tried to go a whole day without touching plastic, you know it's nearly impossible. I'm typing these words to you, getting ready to tell you how detrimental plastic is, on a plastic keyboard. Global production of plastics grew from 15 million metric tons in 1964 to 311million in 2014.[153] In the U.S. in 2017, the plastics industry sold an estimated $432.3 billion in shipments and provided 989,000 jobs in 2017[154] and is expected to grow to $654.38 billion by 2020.[155] Polyethylene terephthalate fiber (PET) is a plastic blended with wool and cotton for cloth and clothing and projected sales by 2035 look like the *Cliff Hangers* mountain on *Price is Right*…they goin' up baby.

In a very eye-opening article by James Bruggers in **Inside Climate**

---

[153] McKinsey & Company: Rethinking the future of plastics
https://www.mckinsey.com/business-functions/sustainability/our-insights/rethinking-the-future-of-plastics
[154] Plastics Industry Association: 2018 Size & Impact
https://www.plasticsindustry.org/sites/default/files/SizeAndImpactReport_Summary.pdf
[155] CISION PR Newswire: Plastics Market Worth $654.38 Billion By 2020: Grand View Research, Inc. https://www.prnewswire.com/news-releases/plastics-market-worth-65438-billion-by-2020-grand-view-research-inc-511720541.html

News, *What's Worrying the Plastics Industry? Your Reaction to All That Waste, for One,* he lays it all out there. While he covers all the best and worst of it, what he said that disturbed me the most is that the industry has soldiers out there whose primary target is fighting legislation trying to curb plastic pollution and to "push back" on more than 400 bills in dozens of states. They are even lobbying state lawmakers to get "pro-plastic" presentations into schools; like you need to encourage a kid that something colorful and shiny is fun. As is the case for most of what's happening to us, this is great for economy and jobs but awful for our planet and the future for our children.

We began in the ocean so let's start there. Plastic is found literally everywhere now; it's in our food[156], it's in the air we breathe[157], it's in the entire world's water[158] and even our poopies[159]. Every piece of plastic you ever touched still exists somewhere. Of all the plastic ever made, only 9% of that was recycled.[160] It didn't disappear; it just became something else made of plastic. So, what do we do with the stuff? Up until recently, we just threw it away. Most of our garbage went into landfills where it broke apart into tiny pieces of plastic. When the rains come, we're back to that nonpoint source pollution; allowing a potential for some of it to wash away into streams, rivers, tributaries and into the ocean.

Out here in the country, many residents still burn their trash. How does that work with regard to plastic? When plastic waste and food waste are

---

[156] National Geographic: You eat thousands of bits of plastic every year
https://www.nationalgeographic.com/environment/2019/06/you-eat-thousands-of-bits-of-plastic-every-year/

[157] The Guardian: Microplastics 'significantly contaminating the air', scientists warn
https://www.theguardian.com/environment/2019/aug/14/microplastics-found-at-profuse-levels-in-snow-from-arctic-to-alps-contamination

[158] Express: Microplastics in water - Why the bottled water you're drinking is not safe from plastic https://www.express.co.uk/news/science/1169570/Microplastics-in-water-bottle-water-microplastics-drinking-water-pollution-health-risks

[159] Forbes: Guess What Is In Your Poop: Plastic, Suggests
https://www.forbes.com/sites/brucelee/2018/10/22/guess-what-is-in-your-poop-plastic/#55300b976a8e

[160] Our World in Data: FAQs on Plastics https://ourworldindata.org/faq-on-plastics

burned, they produce dioxin and furan. These elements, even in small quantities, can cause death. If dioxin is inhaled, it can instantly cause coughing, shortness of breath and dizziness[161] and if everyone does it, even if you are not directly impacted by the fumes the incineration of plastic produces CO_2; that nasty beast that drives global climate change.

And what about recycling? I mean, we cool kids recycle everything the trash man will allow. We frolic down the grocery isles like kids, still grabbing the most colorful and shiny things we can reach and tossing them over our shoulders when they're empty; to be sent to the magical kingdom of guilt-free consumption called recycling. For many years, the U.S. and other wealthy countries sold their plastic waste to China to be recycled into things they would sell inexpensively to other countries (us), and thus it was miraculously *Recycled*. More than 70% of the world's plastic waste went to China, nearly 7 million tons a year. But in 2018, China said, *No mas!* Except they said it in Chinese, which if I typed what they actually said, neither of us would know exactly *what* they said. Where were we? China said, *no more your filthy Americans!* Or something like that. It turns out; much of what we sent them, up to 70% in Cambodia[162], was unusable. Have you ever torn open a cardboard container and peeled away plastic film? That's one of the components to unusable trash. In April 2019, a study revealed:

The impact of the shift in plastic trade to south-east Asian countries has been staggering – contaminated water supplies, crop death, respiratory illness from exposure to burning plastic, and the rise of organized crime abound in areas most exposed to the flood of new imports. These countries and their people are shouldering the economic, social and environmental costs of that pollution, possibly for generations to

---

[161] The Jakarta Post: Burning plastic waste harmful to health
https://www.thejakartapost.com/life/2018/03/31/burning-plastic-waste-harmful-to-health.html
[162] The Guardian: Where does your plastic go? Global investigation reveals America's dirty secret https://www.theguardian.com/us-news/2019/jun/17/recycled-plastic-america-global-crisis

come.¹⁶³

A report in The Guardian tells us that *U.S. waste makes its way across the world and overwhelms the poorest nations.* The report profiles a 60-year-old Vietnamese mother of 7 on the outskirts of Hanoi who lives on and around piles of American plastic. Her job, at a wage around $6.50 a day, is to strip of the material that can't be recycled and sort what *can* be; translucent plastic in one pile and opaque in another.¹⁶⁴ The poorest people in the world are being exploited to process our trash. The equivalent of 68,000 shipping containers worth of the shite was sent to developing countries who mismanage 70% of their own plastic waste. Here it is again: We're pointing fingers about how all the worst problems come from somewhere else, but in truth, it's only because we've sent ours there; hiding the culpability of our mass consumerism cloaked as benevolent business deals to help starving countries to be more like us. It isn't working. The corrupt are still corrupt, the poor are still poor and your plastic Cheetos bag is still a plastic Cheetos bag, it's just a sea turtle's problem now, instead of yours.

Guys, I've put this bit off for more than a day now. I've been pacing back and forth between reports, videos and news stories trying to find the right words to tell you what's happening to the ocean because of our plastic addiction. I don't want to hurt you with horrible images or scare you but like I've said before, we exist because we share the miracle of existence, so we must also share the burden of suffering and not look away; especially when we have caused it. I'm pulling back my shoulders, adjusting my very large but high-set balls and going in.

On a tiny sliver of land in the Pacific, a baby albatross awaits her mother's return; completely dependent on her parents for the food they

---

[163] DISCARDED: COMMUNITIES ON THE FRONTLINES OF THE GLOBAL PLASTIC CRISIS https://wastetradestories.org/wp-content/uploads/2019/04/Discarded-Report-April-22.pdf

[164] The Guardian: Where does your plastic go? Global investigation reveals America's dirty secret https://www.theguardian.com/us-news/2019/jun/17/recycled-plastic-america-global-crisis

forage at sea. Mom and dad take turns flying over hundreds of miles of ocean to hunt food for their precious babies. They skim along the surface scooping up fish eggs, krill and the occasional squid then carry the feast back to hungry offspring where they regurgitate it into the baby's mouth. Wieteke Holthuijzen, a bird researcher stationed on Midway, describes one such occasion:

I remember seeing an adult come back to feed its chick, and it was having a really hard time regurgitating something. I thought the parent was struggling with an unwieldly morsel like a squid. Then it regurgitated a toothbrush — and the chick ate it. It's awful but actually happens here quite a bit.[165]

A 2015 study published by Proceedings of the National Academy of Sciences, United States, (PNAS) found that 90% of seabirds ingest plastic.[166] That's just one species. Our plastic wreaks havoc on coral reefs, fish, and sea turtles. NOAA reports that the chances of corals getting sick go from 4% to 89% in the presence of plastics. Seals and other marine mammals are killed each year after ingesting plastic or getting entangled in it. The Center for Biological Diversity tells us that nearly 700 species of life are adversely affected by plastic in the ocean.

Studies estimate there are now 15–51 trillion pieces of plastic in the world's oceans — from the equator to the poles, from Arctic ice sheets to the sea floor. Not one square mile of surface ocean anywhere on Earth is free of plastic pollution.

Death by plastic is a horrific way to die. The options are death by entanglement, airway/intestinal obstruction, internal laceration or starvation…with a belly full of plastic. No one is safe from it at sea; from whales to tiny seahorses. If you eat seafood, you're eating it too. It is

---

[165] Oceana.org: On a remote island, baby albatrosses suffer from a diet of plastic trash
https://oceana.org/blog/remote-island-baby-albatrosses-suffer-diet-plastic-trash
[166] PNAS: Threat of plastic pollution to seabirds is global, pervasive, and increasing
https://www.pnas.org/content/early/2015/08/27/1502108112

estimated that by 2050, there will be more plastic in the sea than fish.[167] Furthermore, of all the plastic that enters the ocean, we only see 1% of it. Where does it go? A study published in 2014 reported that 99% of ocean trash was unaccounted for.[168] Where did it go? A 2018 study published by *Geochemical Perspectives* thinks they found it.[169] It turns out; it's sinking to the bottom of the sea. The research found that one liter of water from the Mariana Trench contains thousands of tiny plastic pieces. Our plastic rubbish breaks down and breaks down, never going away just becoming its own proliferation like the never-ending, self-replicating gremlin, Gizmo. But there is nothing adorable about *microplastics*. These tiny offenders have permeated the eco-systems down below the sea and are contaminating life forms we have not yet encountered. We don't even know how to replicate the pressure from deep sea environments to try and assess potential impacts.[170] The ocean is the largest Earthling habitat on our planet and now, there's not an inch of it we haven't defiled.

You've probably already heard about the Great Pacific Garbage Patch…no? Well let me tell you: this is a giant, swirling mass of plastic garbage spinning in the sea between Japan, California and Hawaii. It's more than 1.6 million square kilometers in size but sadly it's only a part of North Pacific Gyre (a gyre is a vortex.) Things like wind, temperature and salinity drive ocean currents and presently, there are five major gyres with several patches of swirling garbage within them which are growing in size each day as we dump the equivalent of a garbage truck of plastic every minute into the ocean.[171]

---

[167] Center for Biological Diversity: OCEAN PLASTICS POLLUTION https://www.biologicaldiversity.org/campaigns/ocean_plastics/
[168] PNAS: Plastic debris in the open ocean https://www.pnas.org/content/111/28/10239
[169] Geochemical Perspectives: Microplastics contaminate the deepest part of the world's ocean https://www.geochemicalperspectivesletters.org/article1829#Fig3
[170] National Geographic: Microplastics found to permeate the ocean's deepest points https://www.nationalgeographic.com/environment/2018/12/microplastic-pollution-is-found-in-deep-sea/
[171] World Economic Forum: Every minute, one garbage truck of plastic is dumped into our oceans. This has to stop https://www.weforum.org/agenda/2016/10/every-minute-one-garbage-truck-of-plastic-is-dumped-into-our-oceans/

I know you're probably getting bummed from all the bad news. Don't be. Let this be empowering for you. Realize how very much power you hold in your hands with every single choice you make; every item you buy. Think of that less fortunate soul who is going to end up with your garbage literally at their front door. Many an island home in the most remote regions is being bombarded with our trash. I have one more thing I want to point out then we will move on to a happier place.

I went into great detail with lots of boring and frightening facts about fossil fuels and fracking because I need for you to understand how impossible it is to continue as a fossil fuel society. Maybe you're one of the incredibly fortunate and environmentally aware souls who do everything you can to avoid using petroleum and coal. But very few of us have found a way to live plastic free. You must understand that plastic is a product of petroleum. The U.S. Energy Information Administration states, "Plastics are made from liquid petroleum gases (LPG), natural gas liquids (NGL), and natural gas. LPG are by-products of petroleum refining, and NGL are removed from natural gas before it enters transmission pipelines." It's all connected; that great big ball of our worst habits and addictions. According to the Center for Biological Diversity, the fossil fuel industry plans to increase plastic production by 40% over the next decade. The oil giants are rapidly building petrochemical plants across the United States to turn fracked gas into plastic. This means more toxic air pollution and plastic in our oceans.

My loves, we've made such a mess here. But every single day, in the course of research, I see miraculous things happening with tremendous force and fortitude. We're moving in the right direction. Entire countries are formulating plans to exact change:

- France first banned lightweight plastic bags, then became the first country in the world to ban plastic cups and plates in 2016. As part of its Energy Transition for Green Growth Act, France's plastic ban will now include straws, coffee stirrers, cotton buds, and other

single-use items that use plastic in their design. The ban will take effect on January 1, 2020[172]
- San Diego has banned Styrofoam food and drink containers
- Washington D.C. has banned plastic straws
- EU parliament approved single-use plastic ban
- Canada, the town of Leaf Rapids, Manitoba banned plastic bags
- In the U.S., San Francisco, Seattle, Los Angeles, Portland and Mexico City banned plastic bags
- The states of Hawaii and North Carolina have banned plastic bags
- States in Australia and India have banned plastic bags
- Countries that have banned disposable plastic bags include Italy, China, Bangladesh, many countries in Africa including Rwanda, Kenya, the Congo, and South Africa
- Canada aims to ban single-use plastics by 2021

We are making monumental strides. You can see dozens of these acts of Corporate Social Responsibility (CSR) and good humaning on the parts of countries, states and organizations here:

https://www.nationalgeographic.com/environment/2018/07/ocean-plastic-pollution-solutions/

But please remember, you are the one with all the power. Every single purchase you make, every product you choose and every convenience you sacrifice for the sake of our planet is the right thing to do. The planet will reward you with a world full of beauty and abundance.

---

[172] Hip Paris: France Expands Single Use Plastic Ban for 2020
https://hipparis.com/2018/12/06/france-expands-single-use-plastic-ban-for-2020/

Chapter 7: Species Extinction

By Paul Hollis

Anthropocene Annihilation

A mass extinction can be defined as the rapid death of large numbers of species in a relatively short period of geologic time. Most modern scientists believe these extinctions are due to factors like catastrophic global events or widespread environmental change occurring too rapidly for most species to adapt. During such events, it is estimated that 75% or more of all species perish.

It's generally agreed there have been five documentable mass extinctions that have occurred during the Earth's lifetime. Paleontologists spot notable mass extinctions when species go missing from the global fossil record. "We don't always know what caused them but most had something to do with rapid climate change", says Melbourne Museum paleontologist Rolf Schmidt.[173]

The End-Ordovician mass extinction occurred about 444 million years ago with an estimated loss of 86% of marine species. All of the major animal groups of the Ordovician oceans survived, including trilobites, brachiopods, corals, crinoids and graptolites, but each lost important members. Widespread families of trilobites disappeared and graptolites came close to total extinction.[174] The End-Ordovician extinction is generally attributed to two factors: the first wave of extinction may be related to rapid cooling at the end of the Ordovician Period. This rapid

---

[173] COSMOS – The Five Big Mass Extinctions, https://cosmosmagazine.com/palaeontology/big-five-extinctions
[174] Sam Noble Museum, Oklahoma's Museum of Natural History - End-Ordovician Extinction, https://samnoblemuseum.ou.edu/understanding-extinction/mass-extinctions/end-ordovician-extinction/

cooling could have been triggered by the uplift of several key mountain ranges. The newly exposed silicate rock sucked CO_2 out of the atmosphere, thus quickly chilling the planet. The second phase is widely regarded as having been caused by the sea-level fall associated with glaciation.[175]

Even though continental drift and the creation of mountains is the most accepted cause of this mass extinction, other theories attributed to the End-Ordovician mass extinction are worldwide volcanic eruptions and bursts of gamma radiation from a hyper nova explosion somewhere in our galaxy. Still, it is clear that climate change was ultimately at its root. A major ice age is known to have occurred in the southern hemisphere and climates cooled world-wide. The first wave of extinctions happened as the climate became colder and a second pulse occurred as climates warmed at the end of this ice age.[176]

The Late Devonian mass extinction occurred 375 million years ago, with an estimated species loss of 75%. In researching trilobites from that period, it is known they were the most diverse and abundant of the animals that appeared in the Cambrian explosion 550 million years ago. Their great success was helped by their spiky armor and multifaceted eyes. They survived the first great extinction but were nearly wiped out in the second.[177]

In the Late Devonian, large trees evolved and formed the first forests. As plant life expanded, they used up more CO_2 in photosynthesis. When dead plant material decays, CO_2 is returned to the atmosphere, but some plant material (e.g., leaves) was buried in swamps, lakes and rivers. This buried plant material removes carbon permanently from the atmosphere and often forms coal. When we mine coal and burn it we return CO_2 to

---

[175] Encyclopedia Britannica - Ordovician-Silurian extinction,
https://www.britannica.com/science/Ordovician-Silurian-extinction
[176] Sam Noble Museum – End-Ordovician Extinction
https://samnoblemuseum.ou.edu/understanding-extinction/mass-extinctions/end-ordovician-extinction/
[177] COSMOS – The Five Big Mass Extinctions,
https://cosmosmagazine.com/palaeontology/big-five-extinctions

the atmosphere and warm the planet.[178] As a result, cooling of the planet may have been caused by a drastic drop in greenhouse gases caused by a falling amount of CO_2 in the atmosphere.

The End-Permian mass extinction occurred 251 million years ago with an estimated species loss of an inconceivable 96% of all marine life and 70% of land species. Something killed almost all life on the planet. Less than 5% of the animal species in the seas survived. On land less than a third of the large animal species made it. Nearly all the trees died. Researchers have been able to determine that the End-Permian extinction occurred suddenly during a 31,000-year time period. That is a mere instant in geologic time. The exact cause is still under investigation by scientists but here are a few theories.

First, an enormous asteroid impact is the prime suspect for Gregory Retallack, a geologist at the University of Oregon. The collision would have sent billions of particles into the atmosphere, he explains. They would have spread around the planet then rained down on land and sea. A team of researchers recently found what may be that impact's footprint buried below Australia—a 75-mile-wide crater left by an asteroid more than three miles across. In this case, the short-term effects alone—cold, darkness, and acid rain—would kill plants and photosynthetic plankton, the base of most food chains. Herbivores would starve, as would the carnivores that fed on the plant-eaters.[179]

Many scientists believe that life was already struggling when the putative space rock arrived. Our planet was in the throes of severe volcanism. In a region that is now called Siberia, 1.5 million cubic kilometers of lava flowed from an awesome fissure in the crust (in comparison, Mt. St. Helens unleashed about one cubic kilometer of lava in 1980). Such an

---

[178] Sam Noble Museum – End-Ordovician Extinction
https://samnoblemuseum.ou.edu/understanding-extinction/mass-extinctions/late-devonian-extinctions/
[179] National Geographic - SCIENCE & INNOVATION / REFERENCE The Permian Extinction—When Life Nearly Came to an End
https://www.nationalgeographic.com/science/prehistoric-world/permian-extinction/

eruption would have scorched vast expanses of land, clouded the atmosphere with dust, and released climate-altering greenhouse gases.[180]

Still others are undecided regarding the reason for extinction due to a lack of substantial evidence. "The global ecologic collapse came with a sudden blow, and we cannot see its smoking gun in the sediments that record extinction," Ramezani says. "The key in this paper is the abruptness of the extinction. Any hypothesis that says the extinction was caused by gradual environmental change during the late Permian—all those slow processes, we can rule out. It looks like a sudden punch comes in, and we're still trying to figure out what it meant and what exactly caused it." There is "growing evidence that Earth's major extinction events occur on very short timescales, geologically speaking," says Jonathan Payne, professor of geological sciences and biology at Stanford University. "It is even possible that the main pulse of Permian extinction occurred in just a few centuries. If it turns out to reflect an environmental tipping point within a longer interval of ongoing environmental change, that should make us particularly concerned about potential parallels to global change happening in the world around us right now."[181]

The End-Triassic mass extinction occurred 200 million years ago, with an estimated species loss of 80%. Some scientists say this mass extinction could be a result of climate change and rising sea levels due to too much CO_2 in the atmosphere while others suggest it was because of major Earthquakes along the moving tectonic plate boundaries.[182] Plate boundaries are the edges where two plates meet. Most geologic activities, including volcanoes, earthquakes, and mountain building, take place at

---

[180] NASA SCIENCE - SHARE THE SCIENCE, The Great Dying
https://science.nasa.gov/science-news/science-at-nasa/2002/28jan_extinction
[181] End-Permian extinction, which wiped out most of Earth's species, was instantaneous in geological time
by Jennifer Chu, Massachusetts Institute of Technology - https://phys.org/news/2018-09-end-permian-extinction-earth-species-instantaneous.html
[182] Encyclopedia Britannia: End-Triassic extinction
https://www.britannica.com/science/end-Triassic-extinction

plate boundaries. ... Divergent plate boundaries occur when two plates move away from each other. Convergent plate boundaries occur when two plates move towards each other.[183]

Unfortunately, much of the evidence that might point to the cause has either been destroyed or is concealed deep within the Earth under many layers of rock and therefore remains a subject of endless debate. But world scientists are hard at work examining what little evidence they may find beneath the planet's skin.

Let's examine some other theories. First, one study suggests that during the extinction period, atmospheric CO_2 concentrations were thought to have increased up to four times the pre-extinction levels creating up to 3–4 °C of warming. Increased atmospheric CO_2 concentrations are indicative of ocean acidification suggesting that this may have been a marine extinction mechanism especially in relation to the ocean corals.[184]

Second, volcanism has long been implicated in the extinction, but whether it had a major impact on the planet at the right time has remained unclear. In new research, scientists observed elevated mercury concentrations in extinction-aged rocks from around the world. Because volcanism is the main non-anthropogenic source of mercury in the environment, the findings suggest that (underwater) volcanic activity was likely the main extinction trigger at the end of the Triassic.[185]

These volcanic eruptions are linked to the birth of the central Atlantic Ocean. This "central Atlantic magmatic province" (CAMP) could have released CO_2 and sulfurous compounds into the atmosphere – triggering global warming, acid rain and widespread extinctions on land and at

---

[183] Lumen Physical Geography: Tectonic Plate Boundaries
https://courses.lumenlearning.com/geophysical/chapter/tectonic-plate-boundaries/
[184] Nature Communications - End-Triassic mass extinction started by intrusive CAMP activity https://www.nature.com/articles/ncomms15596
[185] Earth, The Science Behind the Headlines -
https://www.earthmagazine.org/article/volcanism-triggered-end-triassic-extinction

CENTER for BIOLOGICAL DIVERSITY

OUR MISSION: SAVING LIFE ON EARTH

At the Center for Biological Diversity, we believe that the welfare of human beings is deeply linked to nature — to the existence in our world of a vast diversity of wild animals and plants. Because diversity has intrinsic value, and because its loss impoverishes society, we work to secure a future for all species, great and small, hovering on the brink of extinction. We do so through science, law and creative media, with a focus on protecting the lands, waters and climate that species need to survive.

We want those who come after us to inherit a world where the wild is still alive.

The Center's innovation is to systematically and ambitiously use biological data, legal expertise, and the citizen petition provision of the powerful Endangered Species Act to obtain sweeping, legally binding new protections for animals, plants, and their habitat — first in New Mexico, then throughout the Southwest, next through all 11 western states and into other key areas across the country. With each passing year the Center has expanded its territory, which now extends to the protection of species throughout the Pacific and Atlantic oceans and international regions as remote as the North and South poles.

As our range grew, and first tens, then hundreds of species gained protection as a result of our groundbreaking petitions, lawsuits, policy advocacy, and outreach to media, we went from living and working on a shoestring to having offices around the country — from relying on donated time from pro bono attorneys at large firms to building a full-time staff of dozens of prominent environmental lawyers and scientists who work exclusively on our campaigns to save species and the places they need to survive.

We're now fighting a growing number of national and worldwide threats to biodiversity, from the overarching global problems of unsustainable human population and climate change to intensifying domestic sources of species endangerment, such as off-road vehicle excess. Based on our unparalleled record of legal successes — 83% of our lawsuits result in favorable outcomes — we've developed a unique negotiating position with both government agencies and private corporations, enabling us, at times, to secure broad protections for species and habitat without the threat of litigation. We look forward to a future of continued expansion, creativity, and no-holds-barred action on behalf of the world's most critically endangered animals and plants.

Join the team and help us save life on earth (www.biologicaldiversity.org).

sea.[186]

This establishes another theory. The above potential events become mute when the overall timeline is examined. "If you look closely you see there's more subtlety to it," says Lawrence Tanner at Le Moyne College at Syracuse, New York. He and his colleague Spencer Lucas at the New Mexico Museum of Natural History and Science in Albuquerque agree there was ecological upheaval. But they say it took tens of millions of years, so it can't all be (entirely) linked to the CAMP eruptions or climate change. "There is no single extinction at the end of the Triassic," says Lucas. "There's a series of extinctions (over millions of years)."[187]

Finally, the End-Cretaceous mass extinction occurred around 66 million years ago with an estimated species loss of 76%. This is the most famous die-off of the five major mass extinctions to date, ending the reign of the dinosaurs between the Cretaceous and Tertiary periods. Most researchers consider the cause of this mass extinction set in stone so to speak. The case is closed. No, the dinosaurs were not smoking in the school yard where they all contracted simultaneous cancer. Nor did they miss the boat when Noah set sail. You may be thinking of the unicorns there.

End-Cretaceous is the most studied mass extinction event and offers the most evidence of its cause and effects. It is generally accepted today by most scientists that an enormous asteroid or comet collided with Earth at the end of the Cretaceous Period in the Yucatan Peninsula, Mexico, forming what is today called the Chicxulub (CHEEK-sheh-loob) impact crater.[188] Rocks from that age contain traces of the asteroid that struck Earth, generating catastrophic events from global wildfires to toxic gases

---

[186] NewScientist - The mass extinction that might never have happened
https://www.newscientist.com/article/2150939-the-mass-extinction-that-might-never-have-happened/
[187] NewScientist - The mass extinction that might never have happened
https://www.newscientist.com/article/2150939-the-mass-extinction-that-might-never-have-happened/
[188] The Cretaceous Period - UCMP Berkeley
https://ucmp.berkeley.edu/mesozoic/cretaceous/cretaceous.php

to impact debris in the air blocking the sun's light that brought on drastic climate change.

In 1979, a geologist who was studying rock layers between the Cretaceous and Paleogene Periods spotted a thin layer of grey clay separating the two eras. Other scientists found this grey layer all over the world, and tests showed that it contained high concentrations of iridium, an element that is rare on Earth but common in most meteorites, said Betsy Kruk in a class she co-taught on Coursera.org. Also within this layer are indications of "shocked quartz" (basically fused sand) and tiny glass-like globes called tektites that form when rock is suddenly vaporized then immediately cooled, as happens when an extraterrestrial object strikes the Earth with great force.[189]

The Chicxulub crater in the Yucatan dates precisely to this time. The crater site is more than 110 miles in diameter and chemical analysis shows that the sedimentary rock of the area was melted and mixed together by temperatures consistent with the blast impact of an asteroid about 6 miles across striking the Earth at this point.[190]

When the asteroid collided with Earth, its impact triggered shockwaves, massive tsunamis and sent a large cloud of hot rock and dust into the atmosphere, Kruk said. As the super-heated debris fell back to Earth, they started forest fires and increased temperatures. "This rain of hot dust raised global temperatures for hours after the impact, and cooked alive animals that were too large to seek shelter," Kruk said in the class. "Small animals that could shelter underground, underwater, or perhaps in caves or large tree trunks, may have been able to survive this initial heat blast".[191]

That brings us to 2019, where many of us are feeling a bit like the

---

[189] Live Science - Cretaceous Period: Animals, Plants & Extinction Event
https://www.livescience.com/29231-cretaceous-period.html
[190] Live Science - Cretaceous Period: Animals, Plants & Extinction Event
https://www.livescience.com/29231-cretaceous-period.html
[191] Live Science - Cretaceous Period: Animals, Plants & Extinction Event
https://www.livescience.com/29231-cretaceous-period.html

doomed dinosaurs through no fault of our own. But as it turns out, it pretty much is our fault. There still seems to be a tiny bit of debate about whether we're truly in the middle of a sixth extinction. But there is (near) total agreement that the extinctions we're seeing now are our fault.[192]

In fact, there's no longer any question that rising temperatures and increasingly chaotic weather are the work of humanity. There's a 99.9999% chance that humans are the cause of global warming, a February (2019) study reported. That means we've reached the "gold standard" for certainty, a statistical measure typically used in particle physics.[193]

Human beings have existed for just 200,000 years, yet our impact on the planet is so great that scientists around the world are calling for our period in Earth's history to be named 'Anthropocene' – the age of humans. The changes we are now making have exacted a heavy toll on the natural world around us.[194] Even though we are experiencing disturbing signs of Earth's next mass extinction, generally humans are not yet seeing the urgency to change behaviors.

In our present money-based economy of greed, the last few centuries especially have focused on the pursuit of the almighty dollar at the expense of our planet and future human generations. However, if we do not earn money, our personal futures become more imminent and short-lived. It is quite a small corner into which we have painted ourselves. But we continue to seek money to buy more paint.

With 26 billionaires in the United States alone owning more capital wealth that the poorest 4 billion on the planet, it turns out some of us are

---

[192] Business Insider - 15 signs we're in the middle of a 6th mass extinction
https://www.businessinsider.sg/signs-of-6th-mass-extinction-2019-3/
[193] USA Today 04/21/19 - 99.9999% chance humans are causing global warming, and other science-based facts on climate change for Earth Day
https://www.usatoday.com/story/news/nation/2019/04/21/earth-day-2019-climate-change-humans-global-warming-weather-rising-water/3507125002/
[194] Population Matters - Welcome to the Anthropocene
https://populationmatters.org/campaigns/anthropocene

obviously better at making money than others. But here is the economic downside of such a system. It becomes unsustainable when the "haves" and "have-nots" become so polarized that the few remaining "have-somes" required to support both ends of the scale inevitably slip into the unescapable life of the "have-nots". When there is no one left to buy the products the rich produce, the system ultimately fails. This is not an argument against capitalism. It is an argument against the greed that invariably comes with any system that covets money.

So, here we are trapped in a world that extracts the limits of natural resources to continue the pursuit of money and unbridled consumption. But even more important than the financial aspect of the system I described above, is the devastating effect it is having on the condition of our planet. Mother Earth is sick because of the constant pounding, poisoning and pollution of her resources. She's dying in fact, and the life she supports is dying in droves. She has been forced to let them go at a rate of 200 species per day. Please let that sink in. Species. Not 200 animals. Not 200 ants. Not 200 Abyssinian cats. Entire species. Gone forever.[195]

Maintaining a healthy biodiversity is the key to Earth's survival. Each species in an ecosystem has a specific role to play. Because of that strong interconnectedness, the loss of even one species could cause an 'extinction domino effect' to ripple through an ecosystem, causing the entire biological community to collapse. But Earth's biodiversity -- the number of microorganisms, plants, and animals, their genes, and their ecosystems (such as rainforests and grasslands) -- is declining at an alarming rate, even faster than the last mass extinction 65 million years ago.[196]

Countless scientists have documented the warning signs of previous

---

[195] HuffPost: UN Environment Programme: 200 Species Extinct Every Day, Unlike Anything Since Dinosaurs Disappeared 65 Million Years Ago
https://www.huffpost.com/entry/un-environment-programme-_n_684562?
[196] Science Daily - Earth's biodiversity: What do we know and where are we headed?
https://www.sciencedaily.com/releases/2011/03/110310173208.htm

mass extinction events. These same scientists are now agreeing more and more that they are seeing similar signs occurring today. Let's examine some of these indicators.

Insects are dying off at record rates. Roughly 40% of the world's insect species are in decline.[197] Much of this seems to be from habitat loss and human disruption of natural ecosystems. In roughly 50 years, 1,700 species of amphibians, birds, and mammals will face a higher risk of extinction because their natural habitats are shrinking. That is an average rate of extinction over the last century approximately 100 times as high as normal. "There is reason to worry," says lead author of Worldwide Decline of the Entomofauna: A Review of Its Drivers, Francisco Sánchez-Bayo, a researcher at the University of Sydney in Australia. "If we don't stop (the decline), entire ecosystems will collapse due to starvation."

To put this in perspective, declining populations of insects like bees, hoverflies and other pollinators perform a crucial role in fruit, vegetable, and nut production. These little guys provide greater than 20% of pollination services in agricultural production. Without them, it will be impossible to sustain the current human population at sustainable levels.

The study follows several high-profile papers on insect declines that shocked even experts in the field. In October 2017, a group of European researchers found that insect abundance (as measured by biomass) had declined by more than 75% within 63 protected areas in Germany—over the course of just 27 years.

Most of the relevant data comes from Europe, and to a lesser extent the United States, but the rest of the world remains woefully understudied, says David Wagner, an ecologist at the University of Connecticut who wasn't involved in the paper.

---

[197] ScienceDirect - Worldwide decline of the entomofauna: A review of its drivers
https://www.sciencedirect.com/science/article/pii/S0006320718313636

The study found that half of the moth and butterfly species studied are in decline, with one-third threatened with extinction, and the numbers for beetles are almost exactly the same. Meanwhile, nearly half of surveyed bees and ants are threatened. Caddisflies are among the worst off—63% of species are threatened, likely due in part to the fact that they lay their eggs in water, which makes them more vulnerable to pollution and development.[198]

In addition, insects are food sources for bird, fish and mammal species, some of which humans rely on for food. When the smallest link in the food chain is removed, it disrupts the food chain from producer organisms like certain bugs to apex predator species like man and may even eliminate the food chain altogether. This does not bode well for future human generations. See our chapter on Entomology for additional details.

Insects aren't the only creatures taking a severe hit. In the past 50 years, more than 500 amphibian species have declined worldwide — 90 of them going extinct — thanks to a deadly fungal disease called chytridiomycosis that corrodes frog flesh.[199] There are approximately 4,740 species of frogs around the entire world. There are about 90 species of frogs in the United States. Unfortunately, about 120 amphibian species, including frogs, toads and salamanders, have disappeared since 1980. Historically, one species of amphibian would disappear every 250 years.[200]

Half of the total number of animals that once shared the Earth with humans are already gone. In the next 50 years, humans will drive so many mammal species to extinction that Earth's evolutionary diversity won't recover for some 3 million to 5 million years in a best-case

---

[198] National Geographic - Why insect populations are plummeting—and why it matters https://www.nationalgeographic.com/animals/2019/02/why-insect-populations-are-plummeting-and-why-it-matters/
[199] Business Insider - 15 signs we're in the middle of a 6th mass extinction https://www.businessinsider.sg/signs-of-6th-mass-extinction-2019-3/
[200] Defenders of Wildlife – Basic Facts About Frogs https://defenders.org/frogs/basic-facts

scenario to get back to the level of biodiversity we have on Earth today. Returning the planet's biodiversity to the state it was in before modern humans evolved would take even longer – up to 7 million years.[201]

More than 26,500 of the world's species are threatened with extinction, and that number is expected to keep going up. According to the International Union for Conservation of Nature (UICN) Red List, more than 27% of all assessed species on the planet are threatened with extinction. Currently, 40% of the planet's amphibians, 25% of its mammals, and 33% of its coral reefs are threatened. The IUCN predicts that 99.9% of critically endangered species and 67% of endangered species will be lost within the next 100 years.[202]

Our grandchildren may never see a living tiger, rhino, giraffe, elephant or hundreds of other loveable creatures already endangered today. Along with sport egoists who kill for perverted fun, perceived exotic cures and their inevitable paid poachers have decimated the numbers of so many mammal species. Although the penalties are harsh for the "taking" of an endangered species, man has pursued these trophies undeterred.

However, the laws are a bit different when it comes to destroying species habitats that consequently kill entire species at once. In general, the offending companies and groups get a government pass in the name of progress, financial gain or simple paranoia. Clearing of land for logging or livestock pastures have some of the largest negative impacts on habitats.

Areas of high agricultural output tend to have the highest extent of habitat destruction. In the U.S., less than 25% of native vegetation remains in many parts of the East and Midwest. Only 15% of land area remains unmodified by human activities in all of Europe. From the approximately 16 million square kilometers of tropical rainforest habitat

---

[201] Business Insider - 15 signs we're in the middle of a 6th mass extinction
https://www.businessinsider.sg/signs-of-6th-mass-extinction-2019-3/
[202] Business Insider - 15 signs we're in the middle of a 6th mass extinction
https://www.businessinsider.sg/signs-of-6th-mass-extinction-2019-3/

that originally existed worldwide, less than 9 million square kilometers remain today. The current rate of deforestation is 160,000 square kilometers per year, which equates to a loss of approximately 1% of original forest habitat each year. Other forest ecosystems have suffered as much or more destruction as tropical rainforests. Farming and logging have severely disturbed at least 94% of temperate broadleaf forests; and many old growth forest stands have lost more than 98% of their previous area because of human activities. Tropical deciduous dry forests are easier to clear and burn and are more suitable for agriculture and cattle ranching than tropical rainforests; consequently, less than 0.1% of dry forests in Central America's Pacific Coast and less than 8% in Madagascar remain from their original extents.[203]

Evolutionary processes are disrupted when land for logging or livestock is cleared. The species not killed outright from habitat destruction are displaced. These confused animals do not, cannot, just pick up and move somewhere else. Once a habitat is lost or degraded, a species doesn't just wink out of existence: it takes time, often several generations, before a species vanishes for good. A new study in *Science* investigates this process, called "extinction debt", in the Brazilian Amazon and finds that 80-90% of the predicted extinctions of birds, amphibians, and mammals have not yet occurred. But, unless urgent action is taken, the debt will be collected, and these species will vanish for good in the next few decades.[204]

The introduction of foreign species into ecosystems is another big factor in habitat destruction and species extinction. Kudzu is arguably the most famous invasive vegetation in the United States. It was introduced into the southeast U.S. in the last decades of the nineteenth century as a shade plant for new home construction. But kudzu is far too good at providing

---

[203] World Animal Foundation – Habitat Destruction
http://www.worldanimalfoundation.org/articles/article/8948431/181156.htm
[204] Mongabay News & Inspiration from Nature's Frontline -
https://news.mongabay.com/2012/07/still-time-to-save-most-species-in-the-brazilian-amazon/

shade. It smothers other plants and trees under a network of vines and a blanket of leaves, gobbling up the sun and keeping other species from its precious light. Eventually, everything dies.

Another example of invasive species is the Burmese python in the Florida Everglades. These snakes are spread over more than a thousand miles of Southern Florida with an estimated population in the tens of thousands. They eat anything that crosses their path. The swamp food chain becomes their food chain. The entire ecosystem is disrupted and this one may never recover back to a state of ecological balance. A 17-foot, 140-pound python was captured in Big Cypress National Preserve in the Florida Everglades as recently as April 2019.[205] That big guy would instantly destroy my habitat, or I would by beating the bush to get out of there.

One of the most pervasive symptoms of past mass extinctions has been the incredible amounts of CO_2 and heavy metals in the air. The global average atmospheric CO_2 in 2017 was 405.0ppm with a range of uncertainty of plus or minus 0.1 ppm. CO_2 levels today are higher than at any point in at least the past 800,000 years. In fact, the last time the atmospheric CO_2 amounts were this high was more than 3 million years ago, when temperature was 3.6°–5.4°F higher than during the pre-industrial era, and sea level was 15–25 meters (50–80 feet) higher than today.[206] This should sound familiar if you haven't been living in the catacombs beneath Paris for the past decade. With the record level of air-based hydrocarbons, the Earth is once again approaching similar hot temperatures, and with extreme temperatures come glacier melt and rising seas levels.

Since CO_2 is a greenhouse gas, it naturally absorbs heat. Warmed by

---

[205] Washington Post - A 17-foot, 140-pound python was captured in a Florida park. https://www.washingtonpost.com/science/2019/04/07/foot-lb-python-was-captured-fla-state-park-officials-say-its-record/?utm_term=.5d270e20db8c
[206] Climate.gov - Climate Change: Atmospheric Carbon Dioxide https://www.climate.gov/news-features/understanding-climate/climate-change-atmospheric-carbon-dioxide

sunlight, Earth's land and ocean surfaces continuously radiate thermal infrared energy (heat). Unlike oxygen or nitrogen (which make up most of our atmosphere), greenhouse gases absorb that heat and release it gradually over time, like bricks in a fireplace after the fire goes out. Without this natural greenhouse effect, Earth's average annual temperature would be below freezing instead of close to 60°F. But massive increases in greenhouse gases have tipped the Earth's energy budget out of balance, trapping additional heat and raising Earth's average temperature.[207]

Oceans absorb as much as 90% of the heat trapped in the atmosphere. The increases in heat inevitably upset the delicate ecosystem of the oceans. When this happens, marine species and coral reefs die. As one example, researchers in Australia have found that warmer temperatures and water are having a devastating effect on dolphins. Birth rates are down and there is evidence that dolphin lifespans are shortening. They first noticed the effect during a heat wave in 2011. The waters in Shark Bay were 4 degrees warmer than normal – the number of dolphin births fell and the survival rate for some species fell by 12%. Even more worrisome, the effect lasted for at least six years.[208]

Warming oceans (thermal expansion) also lead to sea-level rise. Rising waters are already impacting vulnerable species' habitats. If we don't reduce our greenhouse gas pollution, those levels will rise another 3 or 4 feet on average — and perhaps up to 6.5 feet or more — within this century. Some areas are particularly vulnerable. In the United States, sea levels at hotspots along the East Coast, Gulf of Mexico and northwestern Hawaiian Islands are rising three to four times faster than the global average.[209] This may seem like a small rise in the overall scheme of

---

[207] Climate.gov - Climate Change: Atmospheric Carbon Dioxide
https://www.climate.gov/news-features/understanding-climate/climate-change-atmospheric-carbon-dioxide
[208] Weather Channel - Current Biology Study
https://weather.com/science/environment/video/are-warming-oceans-killing-dolphins
[209] Center for Biological Diversity – Sea Level Rise
https://www.biologicaldiversity.org/campaigns/sea-level_rise/index.html

things until we remember that Louisiana, Florida and much of the eastern coastal regions are at or near sea level.

In fact, Hawaii's iconic Waikiki Beach could soon be underwater as rising sea levels caused by climate change overtake its white sand beaches and bustling city streets. Predicting Honolulu will start experiencing frequent flooding within the next 15 to 20 years, state lawmakers are trying to pass legislation that would spend millions for a coastline protection program aimed at defending the city from regular tidal inundations. The highest tides of recent years have sent seawater flowing across Waikiki Beach and onto roads and sidewalks lining its main thoroughfare, and interactive maps of the Hawaiian Islands show that many parts of the state are expected to be hit by extensive flooding, coastal erosion and loss of infrastructure in coming decades.[210]

Rising seas pose a major risk to our nation's wildlife. The United States is home to 1,383 federally protected threatened and endangered species, many of which depend on coastal and island habitats for survival. Rising seas and increasingly dangerous storm surges threaten to submerge and erode their habitat, and make the groundwater more saline — killing coastal plant communities and ruining drinking water. A groundbreaking report from the Center for Biological Diversity finds that 233 threatened and endangered species in 23 coastal states are at risk from sea-level rise. This means, left unchecked, rising seas threaten the survival of 17% — one out of six — of our nation's federally protected species. The report highlights five at-risk species living in different parts of our coasts: the Hawaiian monk seal, Key deer, loggerhead sea turtle, Delmarva Peninsula fox squirrel and western snowy plover.[211]

Our oceans absorb about a quarter of the CO_2 that humans produce by burning fossil fuels each year, and that's changing their basic chemistry. This is particularly bad for creatures with calcium carbonate in their

---

[210] AP - Waikiki flood concerns spur push for Hawaii shore protection
https://www.apnews.com/9006499dcb6a4978bdb1d14063750550
[211] Center for Biological Diversity – Sea Level Rise
https://www.biologicaldiversity.org/campaigns/sea-level_rise/index.html

shells or skeletons, like mollusks, crabs, and corals. Acidic water makes it harder for them to grow those shells, so many of them are going to have a hard time surviving as our seas change.

Even if you live in the ocean and don't have a shell, bad things could be in store for you. Laboratory studies have shown that increased acidity makes it more difficult for clownfish (a.k.a. Nemo) to sense predators and for sharks (Jaws) to hunt their prey. Scientists estimate that carbon pollution is causing oceans to acidify faster than they have in 300 million years.[212]

You might say, "So, I didn't like those two movies anyway and who cares about sharks and clownfish anyway?" The acidification of our oceans actually affects far more than one predator and one prey. It also affects the oceans biggest predator. Man. In the Gulf of Mexico, nutrient pollution from runoff is combining with carbon pollution in the atmosphere and causing waters to acidify much more quickly than scientists expected, putting the $10 billion fishing industry there at risk. And in Alaska, where half of the seafood caught in this country comes from, rapidly acidifying cold water is endangering 70,000 jobs.

Speaking of direct human involvement in mass extinction, species overfishing adds another dimension to the annihilation of ocean life. Tuna, whales, dolphins, porpoises, salmon, seals and sharks are all endangered from human mismanagement of fishing techniques. Overfishing endangers ocean ecosystems and the billions of people who rely on seafood as a key source of protein. Without sustainable management, our fisheries face collapse — and (humans) face a food crisis.[213]

Not enough studies have been undertaken to determine any immediate health concerns about the seafood we eat because of the growing acidity

---

[212] NRDC - What You Need to Know About Ocean Acidification https://www.nrdc.org/stories/what-you-need-know-about-ocean-acidification
[213] Environmental Defense Fund - Overfishing: The oceans' most serious environmental problem https://www.edf.org/oceans/oceans-most-serious-problem

in the water. Like the effect of warming water temperatures on ocean birth rates and longevity as mentioned above in the Australian study, it seems obvious that increased acidity levels in ocean water has unknown effects on species DNA as fish try to adapt to a new environment. Will fish species become uneatable as acidity increases? It's too soon to tell but the long range effects will certainly come back to haunt mankind with trillions of dollars in increased out-of-pocket costs.

Species that live in fresh water are impacted by a warming planet, too. A 2013 study showed that 82% of native freshwater fish species in California were vulnerable to extinction because of climate change. Most native fish populations are expected to rapidly decline soon, and some will likely be driven to extinction, the study authors said. Fish species that need water colder than 70°F to thrive are especially at risk.[214]

In fact, freshwater fish are the most endangered group of animals on the planet, with more than a third threatened with extinction, according to a report being compiled by British scientists. These scientists have blamed human activities such as overfishing, fresh water pollution and construction for pushing so many species to the brink of extinction.

Dr. William Darwall, manager of the freshwater unit at the IUCN in Cambridge, said: "There are still some big gaps in our knowledge, but of the 5,685 species that have been assessed, 36% of them are threatened. Compared to mammals, where 21% are threatened, and birds, where 12% are threatened, it is clear that fresh water ecosystems are among the most threatened in the world. Sadly, it is also not going to get any better as human need for fresh water, power and food continues to grow and we exploit freshwater environments for these resources."[215]

The rate of freshwater fish extinction on the continent (North America) is

---

[214] PLOS ONE - Climate Change Vulnerability of Native and Alien Freshwater Fishes of California: A Systematic Assessment Approach
https://journals.plos.org/plosone/article?id=10.1371/journal.pone.0063883
[215] The Telegraph - Third of freshwater fish threatened with extinction
https://www.telegraph.co.uk/news/earth/wildlife/8672417/Third-of-freshwater-fish-threatened-with-extinction.html

at least 877 times faster than in the fossil record, where a freshwater fish vanished every 3 million years or so on average. Worldwide, freshwater species are more imperiled than other groups. A scientist in 2009 calculated that freshwater species were currently four to six times more likely to go extinct than their marine and land relatives.[216]

If "business as usual" continues regarding climate change, one in six species living on Earth today is on track to go extinct within the next few generations. Looking at it another way, about 8.7 million species (give or take 1.3 million) is the new, estimated total number of species on Earth. That is a new, estimated total number of species -- the most precise calculation ever offered -- with 6.5 million species found on land and 2.2 million (about 25% of the total) dwelling in the ocean depths.[217] When we consider 200 species are already going extinct each and every day, ALL species now alive on Earth will have perished in just under 140 years on the high side. Allowing for no acceleration as we near the end, humans will watch the last of everything die before we get to the year 2160. That's less than six generations.

Conclusions:

As mentioned, it is widely agreed in scientific circles that Earth will need about 3 million years to regenerate itself after a mass extinction event. Returning the planet's biodiversity to the state it was in before modern humans evolved would take even longer – up to 7 million years. Some paleo biologists take that statistic even further. According to a new study, it (could) take the planet 10 million years or more to recover from a mass extinction event. A study published in the journal Nature Ecology and Evolution reveals that it took around 10 million years for Earth's biodiversity to rebound from the mass extinction that wiped out the

---

[216] Mongabay News & Inspiration from Nature's Frontline - North American freshwater fish going extinct at rate over 800 times the fossil record
https://news.mongabay.com/2012/08/north-american-freshwater-fish-going-extinct-at-rate-over-800-times-the-fossil-record/
[217] Science Daily - How many species on Earth?
https://www.sciencedaily.com/releases/2011/08/110823180459.htm

dinosaurs. "Biodiversity losses won't be replaced for millions of years, and so when you imagine extinctions in coral reef ecosystems, or rain forest ecosystems, or grasslands, or wherever, those places are going to be less diverse essentially forever, as far as humans are concerned," Chris Lowery, paleo biologist and co-author of the study, told Business Insider.[218]

The ultimate triggers of the first five mass extinctions may be somewhat debatable but what remained after the dust settled was the same. Scientists have learned much about triggers and causes of mass extinctions through exhaustive studies described in research paper after research paper. We are seeing the same symptoms today; drastic climate change, high concentrations of CO_2 in the atmosphere, quickly warming temperatures, rising sea levels and sudden accelerating mass species extinction.

The past five years have been the five warmest since record-keeping began in the late 1800s. The Earth has experienced 42 straight years (since 1977) with an above-average global temperature, according to the National Oceanic and Atmospheric Administration. Based on five separate data sets that keep track of the Earth's climate, the global average temperature for the first 10 months of 2018 was about 1.8 degrees above what it was in the late 1800s. That was when industry started to emit large amounts of greenhouse gases into the atmosphere.[219]

Increasing amounts of CO_2 and other gases being released into the atmosphere by industry, transportation and energy production from burning fossil fuels are enhancing what's known as the planet's natural greenhouse effect. CO_2 is the most prevalent among all greenhouse gases produced by human activities, attributed to the burning of fossil fuels.

---

[218] Business Insider - 15 signs we're in the middle of a 6th mass extinction
https://www.businessinsider.sg/signs-of-6th-mass-extinction-2019-3/
[219] USA Today - 99.9999% chance humans are causing global warming, and other science-based facts on climate change for Earth Day
https://www.usatoday.com/story/news/nation/2019/04/21/earth-day-2019-climate-change-humans-global-warming-weather-rising-water/3507125002/

The atmospheric CO_2 level for March 2018 was 411.97 ppm (up 7 ppm in a single year from the 2017 readings) and continues to rise. It has now reached levels in the atmosphere not seen in 3 million years. That's an increase of 46% from just before the Industrial Revolution in the 1800s, when CO_2 levels were around 280 parts per million. Levels began to rise when humans began to burn large amounts of fossil fuels to run factories and heat homes, releasing CO_2 and other greenhouse gasses into the atmosphere.[220]

Another consequence of higher temperatures is the melting of the polar ice caps as we've mentioned. The Earth's glaciers are now losing up to 390 billion tons of ice and snow per year, the study suggests. Global warming has caused over 3 trillion tons of ice to melt from Antarctica in the past quarter-century and tripled ice loss there in the past decade, another study, released in June 2018, said.[221]

There is little doubt that these indications have become serious business and a major concern for mankind. Politicians have been slow to react, hesitant to move from a fossil fuel-based economy because, as established earlier, making money is the end goal in current first-world societies. But we have reached the tipping point where climate change is beginning to cost more than rich industrialists can earn with fossil fuel consumption.

Extreme weather events exacerbated in part by climate change killed almost 250 Americans and cost the nation at least $91 billion in 2018, according to the National Oceanic and Atmospheric Administration. Unusual warmth in the western U.S. in 2018 contributed to a disastrous wildfire season that killed dozens of people. In monetary terms, western

---

[220] USA Today - 99.9999% chance humans are causing global warming, and other science-based facts on climate change for Earth Day
https://www.usatoday.com/story/news/nation/2019/04/21/earth-day-2019-climate-change-humans-global-warming-weather-rising-water/3507125002/
[221] USA Today - 99.9999% chance humans are causing global warming, and other science-based facts on climate change for Earth Day
https://www.usatoday.com/story/news/nation/2019/04/21/earth-day-2019-climate-change-humans-global-warming-weather-rising-water/3507125002/

states endured their costliest wildfire season on record: $24 billion in damage.[222] These are only a few examples.

The warning signs above are well documented for the coming sixth mass extinction but all is not yet lost. Today our planet has a distinct advantage over prior mass extinctions. Humans. Yes, the species that is to blame for our predicament can also be the species that saves the Earth for future generations. We have the technology and the tools, and the timing is right to begin our journey of species salvation. Please see our chapter on *Solutions* for actions humans can take now.

"We are the first generation to feel the effects of climate change and the last generation who can do something about it," says President Barack Obama.

---

[222] USA Today - 99.9999% chance humans are causing global warming, and other science-based facts on climate change for Earth Day
https://www.usatoday.com/story/news/nation/2019/04/21/earth-day-2019-climate-change-humans-global-warming-weather-rising-water/3507125002/

Chapter 8: Climate Change

By Paul Hollis

To put us all on the same page from the start, the term "weather" refers to how the atmosphere behaves in a specific area over a short period of time, usually hours or days. "Climate" refers to general weather patterns over a broad area for a long period of time. Both weather and climate account for qualities like temperature, precipitation, and humidity.[223]

By now most of us either believe we are in the midst of an Earth altering climate transformation, or have strong views about the increasing heat index being just another long, hot summer in a string of long, hot summers. There is no middle ground as we approach 2020. Lines have been drawn, sides taken and arms readied for any assault. But whether you agree or not, like it or not, the world as we know it is undergoing climate changes mankind has never before witnessed.

In the end, it doesn't really matter whether you believe humans are causing long-term effects on the climate or it's just the sun's cycle as some of you have claimed. Whether we call it global warming, climate change, or environmental shifting, at this point in the battle it's a little difficult to deny something is going on with the weather. The summer of 2019 continued the breakneck pace set by the last four years for their pedal to the metal increase of worldwide heat indexes. Each year has been hotter than the one before.

But a few degrees either way are frankly not a stressful enough concept for most of us to care about. Even if the facts are overwhelming. We're all busy getting through life, something we call chasing our dreams, and

---

[223] Food and Climate Change http://www.foodsystemprimer.org/food-production/food-and-climate-change/index.html

there is no time to add a small worry of something we can't control. Besides, we're used to whining about sweating our proverbial butts off in a blast furnace during summer anyway.

History has repeatedly shown us that data and facts alone do not inspire humans to change their beliefs or entice them to act.[224] Increased scientific literacy has no correlation with the acceptance of climate change facts.[225] A growing body of research demonstrated that visualization—the ability to create a mental image of the problem—is the most effective approach for motivating behavior change.[226,227] So, let's put climate change into some relatable terms.

When scientists talk about the impact of climate change, it's often expressed in terms of degrees. But whether the number is 2° or 4°, it seems rather abstract. And not that bad ... after all, who wouldn't like to have a couple of degrees on a shivery winter night? But those numbers are averages. It doesn't mean that the temperatures slide up 2° everywhere all the time. The numbers also tend to be expressed in Centigrade so ... for U.S. folks, that's 3.6 to 7.2 degrees Fahrenheit.[228]

Every living thing on the planet exists in symbiosis within its own species and within the environment around them. It may be easier to

---

[224] Zaval L, Cornwell JFM. Effective education and communication strategies to promote environmental engagement. Eur J Educ. 2017;52: 477–486.
https://onlinelibrary.wiley.com/doi/abs/10.1111/ejed.12252

[225] Kahan DM. Climate-Science Communication and the Measurement Problem. Polit Psychol. Wiley/Blackwell (10.1111); 2015;36: 1–43.
https://onlinelibrary.wiley.com/doi/abs/10.1111/pops.12244

[226] Marx SM, Weber EU, Orlove BS, Leiserowitz A, Krantz DH, Roncoli C, et al. Communication and mental processes: Experiential and analytic processing of uncertain climate information. Glob Environ Chang. Pergamon; 2007;17: 47–58.
https://www.sciencedirect.com/science/article/abs/pii/S0959378006000847?via%3Dihub

[227] Sheppard SRJ. Visualizing Climate Change. Routledge; 2012
https://www.taylorfrancis.com/books/9781849776882

[228] By 2050, Boston will be like Atlanta, Portland like San Antonio, and 22% of Earth ... somewhere else Mark Sumner for Daily Kos
https://www.dailykos.com/stories/2019/7/20/1873175/-By-2050-Boston-will-be-like-Atlanta-Portland-like-San-Antonio-and-22-of-Earth-somewhere-else

understand your own body temperature as it relates to these scientifically predicted temperature rises. Imagine your own temperature permanently jumping 3.6° to 7.2° Fahrenheit and hanging between 102.4° and 106° without any expectation of ever dropping back into a *normal* range. You can panic and cry for help but the nearest doctor literally lives hundreds of lightyears away. That is, if they exist at all, but the existence of alien life is discussion for another day.

Besides, we've learned to ignore these and other beings of science in recent years. They have no idea what they are talking about, right? Nope, you're on your own to figure out how to cure yourself. You can scream and run around in circles until your hair ignites but by then I'm fairly certain we all would wish that we had acted quicker with a little preventative medicine.

Envision events occurring in our own world today. Remember that hair on fire scenario? Think about elongated fire seasons with unprecedented massive and uncontrolled fires in the Pacific Northwest. Do you recall that feeling of nausea that comes with your past feverish temperatures? It's kind of like an unexpected volcano in the pit of your stomach, isn't it? Let's reexamine those foggy, shivering, dizzy feelings caused by a few record-breaking rectal thermometer readings. You only need to track the increase in occurrence and intensity of Midwestern tornadoes and hurricanes coming across the Atlantic to devastate more than two-thirds of our coastline to experience the same feeling. In effect, this is exactly how our planet reacts in the face of climate change. We're just seven billion Sarcoptes scabiei mites grumbling with dissatisfaction on the back of a mangy dog, not realizing we caused the mange in the first place.

Now that you have a frame of reference to understand what's happening to our planet, let's look at what some of Earth's scientists are saying about climate change. In truth, 97% of all climate scientists agree that we are in the midst of cataclysmic climate warming[229] and to make the situation worse, the number of papers rejecting AGW [Anthropogenic, or human-caused, Global Warming] is a miniscule proportion of the published research, with the percentage slightly decreasing over time. Among papers expressing a position on AGW, an overwhelming percentage (97.2% based on self-ratings, 97.1% based on abstract ratings) endorses the scientific consensus on AGW.[230]

Let's look for a moment at what some of these scientists are saying.

Absolutely nothing resembling modern-day global warming has happened on Earth for at least the past 2,000 years, a new study published (July 24, 2019) in *Nature* confirms. Since the birth of Jesus Christ, the climate has sometimes naturally changed—some parts of the world have briefly cooled, and some have briefly warmed—but it has never changed as it's changing now. Never once until the Industrial Revolution did temperatures surge in the same direction everywhere at the same time.

From the planet's perspective, one of the most significant events (during this time) occurred on April 5, 1815, when the Indonesian volcano Mount Tambora began to erupt. Tambora was the largest volcanic eruption since the end of the last Ice Age, one of a series of eruptions that pumped huge amounts of sunlight-reflecting gas into the atmosphere. This gas darkened and chilled summers in Europe. It weakened the monsoons in India and West Africa. It allowed glaciers to advance in the Alps.[231]

On a personal note, I was in Anchorage, Alaska when Mount Spurr erupted in August 1992. I heard the rumble from my apartment, thinking

---

[229] Scientific Consensus: Earth's Climate is Warming https://climate.nasa.gov/scientific-consensus/
[230] Consensus on consensus: a synthesis of consensus estimates on human-caused global warming https://iopscience.iop.org/article/10.1088/1748-9326/11/4/048002
[231] No Climate Event in 2,000 Years Compares to What's Happening Now https://www.theatlantic.com/science/archive/2019/07/why-little-ice-age-doesnt-matter/594517/

it might be distant thunder though the sky that evening was a brilliant crystal blue. Outside, I watched an unbelievable sight approach from the west. An unending black cloud slowly moved over the city, first blocking the sun and turning Alaska's perpetual summer daylight to total darkness. Then the sky began to snow. And snow. And snow. But it wasn't snow; it was volcanic ash silently drifting to Earth. One inch, two inches, maybe three in the end.

It settled everywhere. I mean everywhere, like sand after a day at the beach. Inside locked cars and closed apartment building. Everything I ate had ash in it – for days. Water from the faucet was tinged gray. On the street, there was no demarcation of where to drive even if you thought you had somewhere to go. When the wind picked up the next morning, volcanic ash flew with the breeze, creating near zero visibility conditions. Ash darkened the sky for days and the temperature was off ten degrees.

The ash itself felt like a very fine, almost smooth-to-the-touch powder but under a microscope the ash looked more like snowflakes to me. They were course, gritty snowflakes wanting nothing more than to shred your lungs, eyes and skin. Mount Spurr was a small eruption in comparison but being an eyewitness to Nature's wrath brought new respect for the planet I called home. Here is the official Associated Press release from August 19, 1992:

A volcano erupted today for the second time this summer in southwest Alaska, showering light ash on Anchorage, 80 miles to the east.

The volcano, Mount Spurr, a 11,100-foot peak across Cook Inlet from Anchorage, had been dormant for 39 years until a June 27 eruption.

The ash began raining on Anchorage about 8 P.M. local time, a little more than four hours after the volcano erupted. Within minutes, the authorities had closed the city's airports, stranding hundreds of tourists.

Providence Hospital in Anchorage advised people with respiratory problems to stay indoors. The National Weather Service issued a flash-

flood watch for streams in the area, and the Federal Aviation Administration banned planes from within 25 miles of the volcano, Mount Spurr. [232]

In less than 20 years, millions of people in the United States could be exposed to dangerous "off-the-charts" heat conditions of 127°F or more, a startling new report has found. In 60 years over one-third of the population could be exposed to such conditions, "posing unprecedented health risks," the report says. Parts of Florida and Texas would experience the equivalent in days of at least five months per year on average when the heat index—which includes humidity in its calculations—exceeds 100°F. Most of those days will surpass 105 degrees before the end of the century.[233]

The National Weather Service's (NWS) heat index goes up to 127 degrees Fahrenheit. But in as soon as 20 years, the Southeast, Southern Great Plains, and Midwest will also begin to experience days that are so hot they are "off the charts." A few decades later extreme heat will affect communities in 47 states, said the report, which is also published (on July 24, 2019) as a peer-reviewed study in Environmental Research Communications.[234]

Another study examines the effects of climate change in long range analysis of city pairs for 520 major cities of the world in the year 2050, a mere thirty years from now. They test if their climate in 2050 will resemble more closely to their own current climate conditions or to the current conditions of other cities in different bioclimatic regions. Even

---

[232] AP: Alaskan Volcano Erupts Again, Showering Anchorage With Ash
https://www.nytimes.com/1992/08/19/us/alaskan-volcano-erupts-again-showering-anchorage-with-ash.html

[233] 'Off-the-charts' heat to affect millions in U.S. in coming decades
https://www.nationalgeographic.com/environment/2019/07/extreme-heat-to-affect-millions-of-americans/

[234] Increased frequency of and population exposure to extreme heat index days in the United States during the 21st century
https://www.ucsusa.org/sites/default/files/attach/2019/07/killer-heat-environmental-research-communications-article.pdf

under an optimistic climate scenario (best case scenario), we found that 77% of future cities are very likely to experience a climate that is closer to that of another existing city than to its own current climate. In addition, 22% of cities will experience climate conditions that are not currently experienced by any existing major cities. As a general trend, we found that all the cities tend to shift towards the sub-tropics, with cities from the Northern hemisphere shifting to warmer conditions, (weather) on average (same as currently) ~1000 km south…and cities from the tropics shifting to drier conditions[235].

The study characterized the climate of the world's 520 major cities using 19 climatic variables that reflect the variability in temperature and precipitation regimes for current and future conditions. Future conditions are estimated using an optimistic Representative Concentration Pathway (best case scenario), which considers a stabilization of CO_2 emissions by mid-century[236]. This means the parts per billion of CO_2 in the 2030 atmosphere has been controlled and mitigated by using "effective mitigation policies".

Overall 78% of the 520 Future Cities studied present a climate within the hyper volume representing covered combinations of climate conditions. Therefore, 22% of the Future Cities' climate conditions would disappear from this current climatic domain. As such, 22% of the world's cities are likely to exist in a climatic regime that does not current exist on the planet today. The situation is even more pronounced in the tropics, with 30% of cities experiencing novel climate conditions essentially because the climate will get drier. For example, across Europe, both summers and winters will get warmer, with average increases of 6.3°F and 8.5°F, respectively. These changes would be equivalent to a city shifting ~1,000 km further south towards the subtropics, **under current climate**

---

[235] PLOS ONE: Understanding climate change from a global analysis of city analogues https://journals.plos.org/plosone/article?id=10.1371/journal.pone.0217592
[236] PLOS ONE: Understanding climate change from a global analysis of city analogues https://journals.plos.org/plosone/article?id=10.1371/journal.pone.0217592

conditions.[237] Obviously, temperatures will be higher if we fail to curb CO_2 and other greenhouse emissions.

The new climates in the city comparisons are noteworthy. Madrid's climate in 2050 will be more similar to the current climate in Marrakech than to Madrid's climate today; London will be more similar to Barcelona, Stockholm to Budapest; Moscow to Sofia; Portland to San Antonio, San Francisco to Lisbon, Tokyo to Changsha China, etc. This may not seem so bad to most of you but here is the real problem.

Cities in the tropical regions are predicted to experience smaller changes in average temperature, relative to the higher latitudes. However, shifts in rainfall regimes will dominate the tropical cities. This is characterized by both increases in extreme precipitation events (+5% rainfall wettest month) and, the severity and intensity of droughts (-14% rainfall driest month). With more severe droughts, tropical cities will move towards the subtropics, i.e. towards drier climates. Those 30% of cities in the tropics might also be thought of as cities that could be well on their way to simply being unlivable (cities like Manaus, Libreville, Kuala Lumpur, Jakarta, Rangoon, and Singapore are some examples). And that's without considering sea level rise, or the effect of climate on the food supply. Those people, those billions of people, might well be expected to go elsewhere, dragging with them the heavy baggage of immigration, hunger, and initial financial dependence. And those places they move to will also be experiencing huge changes in their own environments.[238]

Well, you get the idea about what scientists are saying about heat related climate change. Every day there is some new warning in the media. But, let's not point bony fingers at one another. As I mentioned, the damage is ongoing now and increasing. We must understand climate change's

---

[237] PLOS ONE: Understanding climate change from a global analysis of city analogues https://journals.plos.org/plosone/article?id=10.1371/journal.pone.0217592

[238] By 2050, Boston will be like Atlanta, Portland like San Antonio, and 22% of Earth ... somewhere else https://www.dailykos.com/stories/2019/7/20/1873175/-By-2050-Boston-will-be-like-Atlanta-Portland-like-San-Antonio-and-22-of-Earth-somewhere-else

effects on our home – the whole home, not just the basement where we live and the sundeck where we vacation - so we may attack the causes with effective solutions to benefit all species. Please see our chapter on *Solutions* for more detail.

One of the most important consequences of climate change could be its effects on agriculture. The world population is expected to grow to almost 10 billion by 2050. With 3.4 billion more mouths to feed, and the growing desire of the middle class for meat and dairy in developing countries, global demand for food could increase by between 59 and 98%. This means that agriculture around the world needs to step up production and increase yields. But scientists say that the impacts of climate change—higher temperatures, extreme weather, drought, increasing levels of CO_2 and sea level rise—threaten to decrease the quantity and jeopardize the quality of our food supplies.[239]

Although much research has focused on questions of food security, less has been devoted to assessing the wider health impacts of future changes in agricultural production. Agriculture has always been at the mercy of unpredictable weather, but a rapidly changing climate is making agriculture an even more vulnerable enterprise. In some regions, warmer temperatures may increase crop yields. The overall impact of climate change on agriculture, however, is expected to be negative—reducing food supplies and raising food prices.

Many regions already suffering from high rates of hunger and food insecurity, including parts of sub-Saharan Africa and South Asia, are predicted to experience the greatest declines in food production. Elevated levels of atmospheric CO_2 are also expected to lower levels of zinc, iron, and other important nutrients in crops. With changes in rainfall patterns, farmers face dual threats from flooding and drought. Both extremes can destroy crops. Flooding washes away fertile topsoil that farmers depend on for productivity, while droughts dry it out, making it more easily

---

[239] How Climate Change Will Alter Our Food
https://blogs.ei.columbia.edu/2018/07/25/climate-change-food-agriculture/

blown or washed away. Higher temperatures increase crops' water needs, making them even more vulnerable during dry periods.[240]

A recent study of global vegetable and legume production concluded that if greenhouse gas emissions continue on their current trajectory, yields could fall by 35% by 2100 due to water scarcity and increased salinity and ozone.[241]

Another new study found that U.S. production of corn, much of which is used to feed livestock and make biofuel, could be cut in half by a 4°C (7.2°F) increase in global temperatures—which could happen by 2100 if we don't reduce our greenhouse gas emissions. If we limit warming to under 2°C (3.6°F), the goal of the Paris climate accord, U.S. corn production could still decrease by about 18%. Researchers also found that the risk of the world's top four corn exporters (U.S., Brazil, Argentina and Ukraine) suffering simultaneous crop failures of 10% or more is about 7% with a 2°C increase in temperature. If temperatures rise 4°C, the odds shoot up to a staggering 86%.[242]

Eighty% of the world's crops are rain fed, so most farmers depend on the predictable weather agriculture has adapted to in order to produce their crops. However, climate change is altering rainfall patterns around the world. When temperatures rise, the warmer air holds more moisture and can make precipitation more intense. Extreme precipitation events, which are becoming more common, can directly damage crops, resulting in decreased yields.

Flooding resulting from the growing intensity of tropical storms and sea level rise from melting ice and thermal expansion is also likely to increase with climate change, and can drown crops. Because floodwaters can transport sewage, manure or pollutants from roads, farms and lawns,

---

[240] Johns Hopkins: Food and Climate Change http://www.foodsystemprimer.org/food-production/food-and-climate-change/index.html
[241] Columbia University: How Climate Change Will Alter Our Food https://blogs.ei.columbia.edu/2018/07/25/climate-change-food-agriculture/
[242] PNAS: Future warming increases probability of globally synchronized maize production shocks https://www.pnas.org/content/115/26/6644

more pathogens and toxins could find their way into our food.

Hotter weather will lead to faster evaporation, resulting in more droughts and water shortages—so there will be less water for irrigation just when it is needed most. About 10% of the crops grown in the world's major food production regions are irrigated with groundwater that is non-renewable. In other words, aquifers are being drained faster than they're refilling—a problem which will only get worse as the world continues to heat up, says Michael Puma, director of Columbia's Center for Climate Systems Research.

Climate projections show that droughts will become more common in much of the U.S., especially the southwest. In other parts of the world, drought and water shortages are expected to affect the production of rice, which is a staple food for more than half of the people on Earth. During severe drought years, rain fed rice yields have decreased 17 to 40%. In South and Southeast Asia, 23 million hectares of rain fed rice production areas are already subject to water scarcity, and recurring drought affects almost 80% of the rain fed rice growing areas of Africa.

Extreme weather, including heavy storms and drought, can also disrupt food transport. Unless food is stored properly, this could increase the risk of spoilage and contamination and result in more food-borne illness. In 2012, a severe summer drought reduced shipping traffic on the Mississippi River, a major route for transporting crops from the Midwest. The decrease in barge traffic resulted in significant food and economic losses. Flooding that followed in the spring caused additional delays in food transport.

While higher CO_2 levels can stimulate plant growth and increase the amount of carbohydrates the plant produces, this comes at the expense of protein, vitamin and mineral content. Researchers found that plants' protein content will likely decrease significantly if CO_2 levels reach 540 to 960 ppm, which we are projected to reach by 2100. (We are currently at 415 ppm.) Studies show that barley, wheat, potatoes and rice have 6 to 15% lower concentrations of protein when grown at those levels of CO_2.

The protein content of corn and sorghum, however, did not decline significantly.

Moreover, the concentrations of important elements—such as iron, zinc, calcium, magnesium, copper, sulfur, phosphorus and nitrogen—are expected to decrease with more CO_2 in the atmosphere. When CO_2 levels rise, the openings in plant shoots and leaves shrink, so they lose less water. Research suggests that as plants lose water more slowly, their circulation slows down, and they draw in less nitrogen and minerals from the soil. Vitamin B levels in crops may drop as well because nitrogen in plants is critical for producing these vitamins. In one study, rice grown with elevated CO_2 concentrations contained 17% less vitamin B1 (thiamine), 17% less vitamin B2 (riboflavin), 13% less vitamin B5 (pantothenic acid), and 30% less vitamin B9 (folate) than rice grown under current CO_2 levels.[243]

According to a 2011 National Academy of Sciences report, for every degree Celsius that the global thermostat raises, there will be a 5 to 15% decrease in overall crop production. Heat waves, which are expected to become more frequent, make livestock less fertile and more vulnerable to disease. Dairy cows are especially sensitive to heat, so milk production could decline. Parasites and diseases that target livestock thrive in warm, moist conditions. This could result in livestock farmers treating parasites and animal diseases by using more chemicals and veterinary medicines, which might then enter the food chain.[244]

So, climate change is not only bringing us increased heat; it is also the cause of more frequent and intense tornados and hurricanes, volcanic eruptions, Earthquakes, and wild fires. And as we have just seen, our changing world is already hugely impacting our food chain; quantity, quality, and dietary nutrients. Ok, so we may believe we can live through this but it gets seriously scary for most people when it starts to affect our

---

[243] Columbia University: How Climate Change Will Alter Our Food https://blogs.ei.columbia.edu/2018/07/25/climate-change-food-agriculture/
[244] Warming World: Impacts by degree http://dels.nas.edu/resources/static-assets/materials-based-on-reports/booklets/warming_world_final.pdf

pocketbooks. Let's examine exactly how climate change is already beginning to place additional financial burdens on individuals.

Let's look at some current headlines:

Heatwave Ravages European Fields, Sending Wheat Prices Soaring – the next growing season is not expected to be any better as temperatures continue to rise.[245] A bad harvest for farmers directly impacts consumer pricing based on free market supply and demand economics. The less food available means the demand will obviously generate higher market prices. More will be coming out of our pockets to pay for the food humans need to survive.

California Wild Fires Set New Record In 2018 – California suffered $400 billion in damage, according to Accuweather. In addition, it cost the California fire department $1 billion. Both are new records.[246] The cost to fight these forest fires and rebuild what was destroyed is a massive and expensive undertaking directly supported by our federal, state and local tax dollars. Even though government funds are set aside in anticipation of natural disasters, it's never enough because these types of human tragedy are always grossly underestimated. More and more uncontrollable wild fires are occurring because of rising temperatures that cause super dry conditions in vulnerable areas. Taxes will inevitably jump to help cover these costs as climates continue to deteriorate.

Hurricane Michael Insured Losses Hit $6.6 Billion – With more than 15% of claims remaining open as of June 6, 2019, estimated insured losses from last year's Hurricane Michael have topped $6.6 billion, according to new data from the state Office of Insurance Regulation. The number of claims files from the October 10, 2018 category four hurricane has topped 147,325. The overwhelming majority of claims,

---

[245] Heatwave ravages European fields, sending wheat prices soaring
https://www.reuters.com/article/us-europe-wheat-harvest/heatwave-ravages-european-fields-sending-wheat-prices-soaring-idUSKBN1KN0L9

[246] Wildfire Facts, Their Damage, and Effect on the Economy
https://www.thebalance.com/wildfires-economic-impact-4160764

97,484, were filed because of damage to residential property. Overall, 23,194 claims remained open, or about 15.7%.[247] The good news for those affected? Some are collecting federal redress but more are collecting insurance payments to get repairs started. What's the bad news for all others who have insurance coverage across the country? Because of natural disasters like this, insurance premiums become more expensive across the board for all of us to insure our homes, businesses, or other valuable assets in risk-prone areas. Sometimes insurance coverage is even precluded altogether. Either way, our rates are skyrocketing. Insurance companies are not in business to lose money.

Cement Companies Are Starting to Get a $33 Trillion Headache – At first glance this may seem like a strange headline to include with climate change, but investors are now turning on cement producers and other industries, demanding more transparency and action on how they plan to help slow global warming.[248] Cement production accounts for 7% of global carbon emissions. Members of the Institutional Investors Group on Climate change and the Climate Action 100+, a coalition of money managers with more than $33 trillion under management, is asking that European construction-material companies commit to a target of reducing net CO_2 emissions to zero by 2050. This won't be easy for cement companies to achieve but you can be certain costs will be passed along to builders and contracts that will in turn pass these costs along to consumers with a healthy profit included.

U.S. Meat Prices Forecast to Rise - Over the past two decades, food prices have risen 2.6% a year on average but things are about to shift even higher as climate changes progress. While the planet steadily increases temperatures, hot air absorbs more moisture. It rains less, water from lakes and rivers evaporate, and the land dries up. When it does rain,

---

[247] Hurricane Michael insured losses hit $6.6 billion
https://www.wptv.com/weather/hurricane/hurricane-michael-insured-losses-hit-6-6-billion
[248] Cement Companies Are Starting to Get a $33 Trillion Headache
https://www.bloomberg.com/news/articles/2019-07-21/cement-companies-are-starting-to-get-a-33-trillion-headache

the water runs off the land instead of getting absorbed into the water table. That creates floods.[249] None of this has helped meat farmers maintain healthy and viable livestock. As conditions worsen for meat producers, supply decreases, prices increase and those additional costs are passed along to us.

These types of headlines are indicators that life is just going to get much harder than imagined with the added stress of having to make more money because prices are drastically expected to escalate from climate change: food stuff, paper products, tourism, insurance, taxes. We won't be able to depend on those thirty-five cents per hour yearly raises to offset much of these projected increases in this changing economy.

The Nature Climate Change Analysis goes even farther – a report by EPA scientists Jeremy Martinich and Allison Crimmins – examining 22 different climate economic impacts related to health, infrastructure, electricity, water resources, agriculture, and ecosystems. The bottom-line conclusion: by the year 2090, impacts on those 22 economic sectors in just the U.S. would cost about $224 billion more per year…the authors' report comes with an important caveat:

Only a small portion of the impacts of climate change are estimated, and therefore this report captures just a fraction of the potential risks and damages that may be incurred.[250]

And, you can bet we as consumers will be paying those bills.

On September 25, 2019, the third in a series of special reports was released by the Intergovernmental Panel on Climate Change (IPCC) over the past 12 months. IPCC is an arm of the United Nations, dedicated to providing the world with an objective, scientific view of climate change,

---

[249] Why Food Prices Are Rising, Recent Trends, and 2019 Forecast
https://www.thebalance.com/why-are-food-prices-rising-causes-of-food-price-inflation-3306099
[250] EPA: Multi-Model Framework for Quantitative Sectoral Impacts Analysis: A Technical Report for the Fourth National Climate Assessment
https://cfpub.epa.gov/si/si_public_record_Report.cfm?Lab=OAP&dirEntryId=335095

its natural, political and economic impacts and risks, and possible response options.

The study finds that our oceans are getting warmer, the world's ice is melting, and species are moving their habitats far faster than expected. The loss of permanently frozen lands threatens to unleash even more carbon, hastening the climate decline – and it is due in large part to human activity. These conditions have implications for almost every living thing on the planet.

The waters have soaked up more than 90% of the extra heat generated by humans over the past decades, and the rate at which it has taken up this heat has doubled since 1993. The seas were once rising mainly due to thermal expansion – which refers to the way the volume of water expands when it is heated. The extra energy makes the water molecules move around more, causing them to take up more space. But the IPCC says rising water levels are now being driven principally by the melting of Greenland and Antarctica.

"What surprised me the most is the fact that the highest projected sea level rise has been revised upwards and it is now 1.1 metres (that's more than 3.6 feet, folks)," said Dr Jean-Pierre Gattuso, from the CNRS, France's national science agency. "This will have widespread consequences for low lying coasts where almost 700 million people live and it is worrying."[251]

This is a good place for a word about why sea levels are predicted to rise and what will cause the increase. When most people think about rising sea levels, they immediately imagine melting glaciers, but in fact, glacier melt is only one of three reasons. Sea-level rise is governed by processes that alter the volume of water in the global ocean; in order of significance to climate change and increasing temperatures, thermal expansion is the most worrisome, followed by glacier melt, followed by ice sheets losing ice faster than it forms from snowfall.

---

[251] BBCNEWS Climate change: UN panel signals red alert on 'Blue Planet'
https://www.bbc.com/news/science-environment-49817804

We can define thermal expansion, in a nutshell, as simply the expansion of sea water, in this case due to rising temperatures that heat the Earth's surface. Between 1993 and 2018, thermal expansion of the oceans contributed 42% to sea level rise; the melting of temperate glaciers, 21%; Greenland, 15%; and Antarctica, 8%.[252] Climate scientists expect the rate to further accelerate during the 21st century.[253]

Before you say, "yes, but the sea level in my neighborhood is not rising. I'm safe", you need to understand that just as the surface of the Earth is not flat, the surface of the ocean is also not flat—in other words, the sea surface is not changing at the same rate globally. Sea level rise at specific locations may be more or less than the global average due to many local factors: subsidence, upstream flood control, erosion, regional ocean currents, variations in land height, and whether the land is still rebounding from the compressive weight of Ice Age glaciers.[254]

A three-and-a-half-foot rise in sea levels may not sound like a lot unless you are standing in it. Besides, we can just move farther inland, right? Well, here is the rub of that thought. In the United States, almost 40 percent of the population lives in relatively high-population-density coastal areas, where sea level plays a role in flooding, shoreline erosion, and hazards from storms. Globally, eight of the world's 10 largest cities are near a coast, according to the U.N. Atlas of the Oceans. In these urban settings, rising seas threaten infrastructure necessary for local jobs and regional industries. Roads, bridges, subways, water supplies, oil and gas wells, power plants, sewage treatment plants, landfills—virtually all human infrastructure—is at risk from sea level rise.[255]

In conclusion, climate change affects human health and wellbeing through more extreme weather events and wildfires, decreased air

---

[252] Wikipedia: Sea Level Rise
https://en.m.wikipedia.org/wiki/Sea_level_rise
[253] NASA Earth Observatory: Sea Level Rise is Accelerating
https://earthobservatory.nasa.gov/images/91746/sea-level-rise-is-accelerating
[254] NOAA: Is sea level rising? https://oceanservice.noaa.gov/facts/sealevel.html
[255] NOAA: Is sea level rising? https://oceanservice.noaa.gov/facts/sealevel.html

quality, and diseases transmitted by insects, food, and water. Climate disruptions to agriculture have been increasing and are projected to become more severe over this century, a trend that would diminish the security of America's food supply. Surface and groundwater supplies in some regions are already stressed, and water quality is diminishing in many areas, in part due to increasing sediment and contaminant concentrations after heavy downpours.[256]

Those who will be most affected by the impact of climate change are often those least responsible for it. Oxfam estimates that the poorest half of the global population is responsible for 10% of global emissions, while the richest 10% of humanity are responsible for 51%. Climate change is transforming the operational context for every organization on the planet. As well as its direct physical impacts, it is affecting where people live, their capacity to work, feed and nourish themselves, and relationships between people and powers.[257] But individuals who are most vulnerable to environmental changes will be its initial targets:

- The Poor - People who live in poverty may have a difficult time coping with changes. These people have limited financial resources to cope with heat, relocate or evacuate, or respond to increases in the cost of food.
- The Elderly - Older residents make up a larger share of the population in warmer areas of the United States. These areas will likely experience higher temperatures, tropical storms, or extended droughts in the future. The share of the U.S. population composed of adults over age 65 is also projected to grow from 13% in 2010 to 20% by 2050.
- The Young – Our children are another sensitive age group, since their immune system and other bodily systems are still developing and they rely on others to care for them in disaster situations.

---

[256] Climate Change Impacts on Society https://www.globalchange.gov/climate-change/impacts-society
[257] Impacts of climate change https://www.thefuturescentre.org/trend-card/environmental-impacts-climate-change

- Indigenous Peoples – Native cultures relying on surrounding environment and natural resources for food, cultural practices, and income.
- Urban Dwellers - Heat waves are amplified in cities because cities absorb more heat during the day than suburban and rural areas.

So, let's bring this home. The United States has one of the largest populations in the developed world, and is the only developed nation experiencing significant population growth. Its population may double before the end of the century. Its (current) 300+ million inhabitants produce greenhouse gases at a per-capita rate that is more than double that of Europe, five times the global average, and more than 10 times the average of developing nations. The U.S. greenhouse gas contribution is driven by a disastrous combination of high population, significant growth, and massive (and rising) consumption levels, and thus far, lack of political will to end our fossil-fuel addiction.

More than half of the U.S. population now lives in car-dependent suburbs. Cumulatively, we drive 3 trillion miles each year. The average miles traveled per capita is increasing rapidly, and the transportation sector now accounts for one-third of all U.S. carbon emissions.[258]

With all of this said, we are certainly not the worst polluters but none of us are doing the world any good. Because I believe most humans are not innately empathetic or even sympathetic to the plight of those outside their tiny circles of perceived influence, I've tried to focus on the very thing we all understand – the cost to humanity in terms of pocket change. Make no mistake about it. Climate change is going to cost *us* billions; already costing us billions. Did you notice the key word in that sentence? I don't mean the government or big business or the invisible guy in the sky. *Us.* The money is coming out of our pockets, either indirectly through taxes or directly through higher costs for everything that comes

---

[258] Center for Biological Diversity: HUMAN POPULATION GROWTH AND CLIMATE CHANGE
https://www.biologicaldiversity.org/programs/population_and_sustainability/climate/

to mind.

So, what can *I* do? I see that question floating in the air around us as we close this chapter. The worst we can do is do nothing. While progress on decarbonization has accelerated in recent years, it is clear that the pace of change is inadequate. In the words of UN Secretary General Antonio Guterres, "climate change is moving faster than we are".[259]

Please refer to our chapter on *Solutions* for actions we can take right now.

---

[259] Impacts of climate change
https://www.thefuturescentre.org/trend-card/environmental-impacts-climate-change

Chapter 9: My Earth Ministry

By Pamela Dawn

> *All religions are a sliver of the truth.*
> *All God's are a sliver of the One.*
> *All souls are a sliver of the same being.*
> *And all Earthlings are a sliver of one Mother...Life.*
> *See yourself in all life around you.*

The Gaia hypothesis was constructed by a chemist, James Lovelock and co-developed by a microbiologist named Lynn Margulis in the 70s. It has also been called the Gaia theory or Gaia principle. The theory proposes that all living organisms interact with their inorganic surroundings on Earth to form a synergistic and self-regulating, complex system that helps to maintain and perpetuate the conditions for life on the planet.[260]

While there are theories, hypotheses and a great deal of speculation regarding planet Earth, to me, she is Mother of all life. And since it appears that the bacteria which hitched a ride on a meteor bringing life to our planet came from Mars, I guess that would make Mars good ole Dad.

No matter where your philosophical, moral or religious beliefs position you, this message is about *my* experiential knowledge, where it came from and how my brain interprets it.

Do you know that all we know of Socrates comes from his student Plato and contemporary evidence of his life? We find nothing of his pen, only his mind. Widely believed to be the wisest man to have lived, he believed written words were a limitation of knowledge. I love this

---

[260] Wikipedia: Gaia Hypothesis https://en.wikipedia.org/wiki/Gaia_hypothesis

human, even though he died thousands of years ago and what we know is second-hand; albeit from several sources. But, if we gather all the words written about him, a single voice sings out loud and clear, and that voice is not the same voice as the writer; most often his student Plato. He is real enough for me. And I, like the ghost of my teacher Socrates, know what I know now, but you may teach me something that completely alters my relative truth. In a dialogue known as The Phaedrus, he said,

I cannot help feeling, Phaedrus, that writing is unfortunately like painting; for the creations of the painter have the attitude of life, and yet if you ask them a question, they preserve a solemn silence. And the same may be said of speeches. You would imagine that they had intelligence, but if you want to know anything and put a question to one of them, the speaker always gives one unvarying answer. And when they have been once written down they are tumbled about anywhere among those who may or may not understand them, and know not to whom they should reply, to whom not: and, if they are maltreated or abused, they have no parent to protect them; and they cannot protect or defend themselves.

Reading the words in this, our book, you are reading what we know to be true now. But those dead words will continue to say the same thing over and over, never learning to live. If you want to know what *I* believe now, put the book down and ask me. So, why write a book at all? Simply put, a book is a starting place for true "Socratic dialogue".

I believe our planet is a living entity, a being, possibly sentient. Scientists signed a document called The Cambridge Declaration on Consciousness on July 7, 2012, including Stephen Hawking, attesting to the fact that animals are sentient and self-aware:

Convergent evidence indicates that non-human animals have the neuroanatomical, neurochemical, and neurophysiological substrates of conscious states along with the capacity to exhibit intentional behaviors. Consequently, the weight of evidence indicates that humans are not unique in possessing the neurological substrates that generate consciousness. Non-human animals, including all mammals and birds, and many other creatures, including octopuses, also possess these neurological substrates.[261]

There have been recent discoveries about the biology of animals like the orca and elephant which back it up. Their limbic system, the place for familial bonding and emotion, is more developed and significantly bigger than a human's. We have governed this planet by myths of an invisible man in the sky for millennia, why would it be a stretch to believe his wife inhabits this planet? Okay, don't get huffy, I'm not dissing your religion. I believe we all have the right to our beliefs and are entitled to live by them. What I don't see as an entitlement of higher thinking is the right to force our convictions on unsuspecting others who are only trying to enjoy a nice cow latte or pure cruelty-free tofu burger. Share your knowledge when it is welcome, but: Live and let the fuck alone.

When I returned to college the second time, to Lord Fairfax Community College, I switched my major from international business to Philosophy, with an emphasis on religion. I akin religion to *those* vegans who mung bean-thump the facts and attack we meat-eaters while having all the best intentions to do good. It's great if that's your chocolate happy place. It just sucks to not be you when you're all fired up and judgy. "Wait, was she talking to vegans or we Christians? That pisses me off!" See what I mean?

Still here? Then you're ready to hear that this is what I believe: She is a biosphere and every living entity, every system of her mechanism and every life on her *is* her; including you. We are all connected by our source and when we damage a fragile part of this intricate system, we damage ourselves. Although this is something I came to on my own in my search for God in my early thirties, I was overjoyed to find the *Gaia Hypothesis*.[262] As my favorite World Religion professor Rabbi Richman

---

[261] The Cambridge Declaration on Consciousness
http://fcmconference.org/img/CambridgeDeclarationOnConsciousness.pdf
[262] Harvard.edu: Gaia hypothesis
https://courses.seas.harvard.edu/climate/eli/Courses/EPS281r/Sources/Gaia/Gaia-hypothesis-wikipedia.pdf

said, "No one's been original since Adam." I'm giggling. I quote him often. My favorite quote of his came with a story. You'll need to imagine a 150-year-old rabbi with a Yiddish accent.

When I was feeling better than someone else, my fathah would say, "If God can stand 'em, who 'uh you?"

Here's the story: The first few days in his class were torture. His voice was calm, low and a lovely lullaby. Having three kids at home, two still in diapers, it was hard enough to stay awake through lectures, but that voice… Until one day, he stopped speaking for a moment, as he often did, and with a faraway stare said, "Nine could be fakers, one could be real. You don't know the difference, give to all who ask." I was enchanted. I realized this was happening every time he paused. He would get that look as if he was watching a magnificent play unfold on a stage only he could see. Then one of those gems would pop out of his mouth. That gave me such incentive to hang on his every word; just to wait for the payoff quote. By the end of the semester I compiled all the quotes, had them bound and called it, The Wisdom of a Richman, and then gave it to him as a thank you at the end of the semester. I sat with him while he read it. It was my *Tuesdays with Morrie* moment and one of my favorite memories. He laughed, gave commentary and at the end, shed a tear. The next time I saw him, he said his adult children loved it and kept adding quotes to it. He died the following year and my heart is still broken. I like to think he's here now, on that stage only I can see, scoffing at my potty mouth and giggling at my candor. I wonder how he felt about the Gaia Hypothesis, or if he ever read it? I asked him out loud after a couple of glasses of wine one night and my mind heard, *"What? I'm gonna give you the answers?"*

My second favorite professor and a mentor who had a profound impact on the course of my life was Hal McMullen, and he was the one who opened my mind to prepare it for such grandiose ideas as Gaia. He was a true Socratic teacher and walked the path *with* students to allow them to find their truth. He was the dean of philosophy and his passion, like mine at the time, was religion. We spent many hours together and I will never forget the time I shared my cosmological theory of God with him. When I asked him to what religion I belong, he said, "You are a religious Maverick, my dear. No such religion exists. Perhaps you will create one." I am smiling now to remember that moment but I am sitting here firmly grasped in a definitive position: I will not create a new religion. Religion has only served to divide us and I seek to unite us all, although, I *have* since become an ordained minister. Since my life has become a practice of many religions the only logical choice for ordination was Universal Life. Yes, it is a single institution from which anyone with a desire can become ordained, but take it very seriously as I live my life in service. In order to serve in certain places, it is often necessary to have credentials and was, therefore, a necessary tedium for me.

Oh, the places I have served…the lessons I have learned. After struggling seven years with the generalized dystonia and having my life restored, I heard the call to serve. The next question was, serve whom and in what way? I tried on my service hat by living three months in a homeless shelter. Yes, it was as bad as you can imagine. But the people I served changed me.

We were required to attend a bible study class every Wednesday and the man who ran it was a giant, gruff, ex-wrestler. He snarled and barked, "Every one of you will memorize our weekly scripture and recite it to the class, every week." And we did. When it was Mr. Johnson's turn to recite, he sat quietly and said nothing. The teacher growled and demanded recitation. Mr. Johnson was a strong man too. He was like a deep black mountain with eyes full of pain and hard living. He said nothing. Before the next onslaught of venom could fly, a young man spoke up in Mr. Johnson's defense, "Sir, he don't know how to read." My heart sank. We were all hoping for compassion to finally rise up in the man *serving us,* but he grumbled, "That's no excuse. You find a way to learn the next one."

When the class was over, I pulled Mr. Johnson to the side and asked, "Mr. Johnson, would you like to learn to read?" His eyes smiled and he said, "I always did." The next day after lunch, Mr. Johnson walked with me to a nearby general store and I purchased a first reader's tablet and a pack of pencils. As soon as we got back to the shelter, we sat under a big tree and he learned his first letters. Every day for the next three months, after lunch we sat under the tree and practiced. I loved watching him write each letter. He was like a master craftsman, an artist as he slowly and with type-written precision, drew perfectly straight lines and circles. One day, when we had reached "S," he asked, "Do you know how to spell Michael?" I said, yes, and recited each letter as he drew them each with razor focus and exactitude. "Do you know how to spell Anthony?" He continued on the same line as Michael. "Do you know how to spell Johnson?" On the next line, he wrote, "Johnson," then he reached in his back pocket, pulled a driver's license from his wallet and lay it down below the three meticulous words. I single tear rolled like a river down his beautiful ebony cheek. He turned to me and said, "That's my name. I never thought I would write that." Even now, I'm blubbering like a baby as I write this. It was one of the most meaningful experiences of my life. Mr. Johnson and I became the best of friends. I promised him that when I moved on in my life, I would bring him with me. As it turned out, he was moved to a rehab facility and I could never find him again. My heart still breaks for him now and again.

Photos Are That Actual Day

Though few experiences in my life would be as meaningful as my friendship with Mr. Johnson, there were others at the shelter who taught me how to live a thankful life. There was the young man who stood in the street yelling profanities to the *Ghost Configuration* in the sky, as he called it. He was so lost that he lived in urine and feces-soaked clothes. The look on his face when I hugged him told me he was in there somewhere.

One of my favorite characters was Brenda. She was a lady from the streets and had survived a life few of us could even imagine. She was hard and tough as a prizefighter and could have been 25 or 45. I remember one afternoon when a newcomer got on the wrong side of her. From inside, I heard Brenda getting loud and I bolted out the door. People didn't understand her and I imagine she had lived most of her life misunderstood.

The frail, blonde newcomer was coming off the drug *spice* and in her agitation, spouted something hateful to Brenda. Brenda and I had bonded over the course of our few weeks together and I knew we were good the day she said, "I like you girl. You real." But on one afternoon, from my time in private security, I recognized Brenda's posture and things were about to go very badly for spice-girl. I positioned my body between Brenda and spice-girl with my back pressed against Brenda and slightly wrapped an arm behind me and around her. I gradually moved her back a bit at a time and could feel the heat from her body as she tensed and headed into a frenzy with her words. I looked dead in the eye of spice-girl who continued spouting venom and I said, "You are heading for grievous injury, you need to shut your mouth now. She's from the streets and you're about to get the beating of your life." Thankfully, spice girl heard the alarm in my tone. She dropped her eyes and shut her mouth long enough for me and Brenda to take a walk. In a rare moment of vulnerability, she said, "I didn't wanna hurt that girl but damn," and then she bumped my shoulder with hers. The gesture was the most beautiful *thank you* ever.

And then, there was the lovely tall gentleman with turrets. When the shelter congregation attended church, we women usually entered church last. We came in the front door, facing the congregation and he sat in the back. It was almost like a tick; as soon as I walked in he shouted, "Pamela!" in a high pitched chirp. Of course, I'd giggle and wave which probably made it worse because for the rest of the sermon, every now and then he'd chirp, "Pamela! Pamela!" It was a welcome moment of joviality that almost took the sting out of the words the preacher pounded into us: "You are here because you have worn out your welcome with every other person you've met." Or his classic, "God loves you and I'm tryin'." But the injustices I witnessed are for another book, another time.

I had begun to feel that ugly place starting to pull me down and in, like a leaf down a sewer drain, and it was time to move on. I landed in Florida, just down the road from my parents, and realized they needed my service more than strangers. So, I spent my dad's last four years on Earth playing the role of handyman and it was glorious. My Papa was an original *John Wayne* and created me in his image. He never let me feel limited by my gender and taught me to fix cars, plumbing and all manner of stuff that breaks. I'm an original *Joan Wayne* now and I miss him every day. Where were we? Ah yes, my call to serve.

I decided then, I would serve where I stood. I got an online ordination, to make it easier in the event that I happened upon someone in need of my particular education in world religions. I could administer last rights in several traditions, bless new babies, marry those brave enough to play that lottery and pray in all the world's traditions. So, this is the way I minister.

Once in a laundromat, I met two older men struggling to figure out the soap machine. After the soap lesson, I invaded their space with my, "Hi, I'm Pamela" shtick. One was friendlier than the other and an hour into our conversation he revealed that he had recently been released from maximum security prison, where he had been since his early twenties. I must have been the first person he told because he had that look, almost a wince waiting for me to either ask what he did or move uncomfortably away from him. I did neither. It didn't matter what he did; he paid the price. We exchanged phone numbers and over the course of the next few months, he began to consider me his friend and even called me Pastor Pamela.

He became ill and was hospitalized so I went and sat by his side. I made it a regular habit to uncover his feet and wash them with a cool cloth before I left him. I remembered that small act making me feel brand new when my mom washed my feet when I was sick as a kid. One afternoon, he was out for a procedure when I came to check on him. I was bored so my Attention Deficit Disorder (ADD) and I snooped around the room for something to read. I was reading instructions left by the charge nurse and noticed his name at the top. The name he had given me was actually his middle name. It made me curious enough to google him when I got home, as I am ever the cautious and often reluctant humanist. What I read sickened me. I mentioned to you the source of my PTSD, the scars of unspeakable rapes, torture and beatings burst open as I read. What I read was, this man was a violent rapist and was held in a maximum-security prison for the criminally insane for the violence he inflicted on his victim. This was my lesson. This was my trial since his was over. I raged, I cried and pounded my fists into my mattress. How could I continue to serve this man when I know intimately the darkness he attached to another innocent sister? Because it was the right thing to do.

The next day, I walked in his room, filled a basin with water, prepared a soapy washcloth, and proceeded to wash his feet more gently than I had ever done before. I wasn't speaking to him this time, even though he repeatedly did say, "Thank you, you don't have to do that." I remained silent and the only words were spoken by the tears pouring down my face and the trembling of my hands. By the time I was finished, I think he figured it out. As I dried his feet and gently tucked the sheet and blanket back over his clean feet, I said, "I was brutalized and raped many years ago. It changed me. I am strong now but that sixteen-year-old girl died a brutal death and I carry her inside me still. I wish you had told me your name, and then I wouldn't have felt like you were being devious when I learned it."

He said all the things one would expect like, "I was afraid you wouldn't be my friend...I was ashamed...I'm trying to live a good life now." He was right. He had paid for his crime and deserved a new start. But I was, and *am* so fucking mad because we victims never, ever, ever get that chance. We don't walk away from our wounds we just cover them over with lovely philosophies evolved on the dynamics of human behavior and we utter cliché's like, "it made me stronger." The truth is...it did, I am and life, generally, is good. But I still get goddamn pissed when another victim falls and I will never be unwounded. I've come to this conclusion:

The greatest suffering we humans endure is inflicted by fear. But evolution has afforded us some relief. I know by experience, that when abuse reaches the length of its hours and the peak of its brutality, the soul steps away leaving only the body to suffer the indignation. For some, it can't bear to return and the body is left to live out the years behind motionless eyes; always gazing downward. But when the soul returns to an abused body, that creates a new form of human possessed of the courage to run towards the sound of danger, to stop it happening to another.

Elwood and I were never as close again. I served him, I like to think I forgave him; but I am a protector, a defender of the voiceless. Inside, I will always be raging against predators. At any given moment, I have at least one weapon on my body and several more within my reach. Clearly, I lack that certain quality necessary to forgive unconditionally. I'm working on it.

When I told you I wanted to open myself up, to expose myself as an offering to gain your trust; what follows is my underbelly. I want to tell you how I ended up in Religious Studies to begin with. It's a painful story and hard to tell, but if we're going to develop trust on these pages, it's something you should know about me and how this human that has never existed before came to be.

My relationship with religion has been tumultuous, at best; a love-hate thing. I have been on a quest to find my religion since I was 8 years old. My grandfather gave me my very own bible and told me to read it before I saw him again the next year. I did, much to his surprise. I understood some of it, was terrified by most of it, but there was one repeated phrase found several times throughout the Bible, including Genesis and Corinthians, that captivated my full attention: Third Heaven. I suffered from OCD as a child so this was probably some weird-kid obsession, as I became driven to find its meaning.

That summer, I went to church with four different friends, looking for answers. Keep in mind, I was eight. I know, totally weird. But one of my friends took me to a charismatic tent revival. My mom had taken my sister and me to many different services, all denominations in the military chapel, but none of them prepared me for speaking in tongues and literal holy rollers. I was fascinated and mortified all at the same time. When the preacher stopped preaching and said, "Are you ready to live a Godly life in the name of Jesus? Come to the altar and pray for Jesus to forgive your sins and accept you into the kingdom of God."

Holy shit! Yes please! I was a...we'll say, precocious kid. I had lots to be forgiven for; I'd better get up there. And so I did. That line was for you Auntie M. Are we allowed to have private jokes in books? My book, my rules? Rude Pamela. I'll share. When I was really little, like 6, we periodically lived with my grandparents and my aunts who were more like big sisters to me when my dad went to war. They loved to play word games and tried with great frequency to explain it to me but having very poor grammar skills at the time...okay, I see what's goin' on here. Now I know why you let me play. Ha ha. So, there I was in a room full of teenagers in the early '70s who found it terribly amusing when every sentence I played began with, "and so she did." You had to be there. I think it was funnier live because I had a mathive lithp. Anyway, back to the revival. There was laying on of hands, on top of heads, *schmack* against foreheads and eventually hands on my head: Weird, but okay. Then lots of yelling and praying and after it was all over, the preacher wanted to meet with me and my friend to make sure I understood the choice I had made.

Way back then, there was no ADD. We super hyper kids were diagnosed with all manner of conditions and mine was called hyper-kinesis. I was a ROWDY mofo. Add a high IQ to the mix and, well, I'll never understand how my mom didn't end up a drug addict. So, I fidgeted, wiggled, scrunched up my forehead the whole time he was explaining all the moral vicissitudes that would now land me in Hell. Finally, there was a break in the rant and I dove in with, "What's the third heaven?" He looked stunned. "What's that?" I repeated myself and flipped open my bible to several of the references. Straight on, deer-in-the-headlights:

"Now, that's just sumthin' you don't need to worry about."

"But I thought there was only one heaven."

"Little girl, I'm tellin' you, there's things you'll understand about the word of God when you're older. That's not somethin' you need to know right now. Do you want to go to heaven and be with Jesus one day?"

"Yes sir, but which heaven will I go to?"

That was it. His *are-you-fucking-kidding-me* meter done broke. He grabbed my bible, slammed it shut, threw it sliding down the long table and shouted to my friend's mom, "You need ta take this little girl back where you got 'er."

---

10 years later...The first time I went to college, I was a struggling single mom. I was only 18, trying to carry a full load at Merced College and still feed and find care for my 10-month-old baby. I had chosen to stay in my hometown of Winton, California, to be near my big sister who saved my life from that abusive two-and-a-half-year marriage. After the hell I had survived, the last thing on my mind was love. But I met a very determined and charming young man named Morgan who also happened to be class president and he was convinced I was the first lady of his dreams. I was sports editor of our school paper and his office was inside our workroom. We were a motley bunch of nerdy writers living on the fringes of freedom and social liberation, with no clue how to blend. But Morgan was the cool kid. I adored him. He was funny, brilliant and so very set on becoming president one day. Once, we stayed up talking until the sun came up and I told him how I had never stopped looking for the answer to a single question and the traumatic memory of my first attempt to have an authoritative answer. So, that night, when he told me he held the priesthood in his religion, guess what I immediately slung at him? But this priest had the answer. He went on for an hour explaining the Church of Jesus Christ of Latter-Day Saints (LDS) doctrine and how the heavens are divided. I was definitely intrigued.

We became close friends. As I look back now, I'm relatively certain Morgan was behind my nomination to be Homecoming Queen. But, for those of us in the journalism club, it was our ticket to the big show. My fellow grammar enthusiasts thought for certain even a nomination would make our club more socially palatable, but I was the geek who would have to face the cool kids, find a dress while having the style sense of a lumberjack and worst...find a date. Morgan came to the rescue.

I didn't win but the night of the gala, he picked me up in his dad's Mercedes. I in my gown and he in his tux, he took me to dinner...at

McDonalds. I'm laughing so hard right now. We were early Millennials, basking in the irony of it. We left the dance early and he drove...and drove...and drove and told me I'd love where we were headed. About the time we were driving over the Grapevine, I realized we were heading to my favorite city at that time San Francisco. But what happened next was all Morgan. He parked the car and took me by the hand to an overlook. We were looking out at Alcatraz. As the sun began to rise behind Alcatraz, (I'm laughing again) Morgan bent to one knee and proposed. Pure Millennial irony, right? Of course, I accepted. I suppose I should have seen the dark omen of what then, I thought was terribly romantic.

But life was about to change my circumstances. My sister went back to her boyfriend and I could no longer survive on my own. So, baby Misha and I would be moving home. As happens so many times in our lives, aware or unaware of it happening, a series of coincidences gathered the main players in a significant chapter of my life in Washington D.C. that year. I already mentioned, Papa was working for Honeywell on the FA18 program and was transferred to Reston, a suburb of Washington D.C. Morgan was determined not to lose us, so he accepted an internship with Congressman Lehman. Within weeks of our moving there, Morgan rented a room from a church member and joined us in Northern Virginia.

He still had a few tricks up his sleeve, like; he re-proposed on the River Boat Dandy, and then to consummate our engagement, sealed it with a kiss on the steps of Congress. Keeping in mind that he was Mormon and I was very young, living with PTSD from multiple physical and sexual assaults, there had been no hanky-panky and a kiss was all we needed. It was a nice kiss. There were no fireworks but he made me feel safe, I loved his company and a life with Morgan was everything my parents had hoped for.

I was fortunate to have two remarkable men to call Dad; okay, one Dad and one "Papa." Oh, Papa...he remains my hero. He was 6'4" and when we met him, he was working on the simulators with my Dad and was a cop in Merced at night. But that night in Reston, the night Morgan took me home from our official engagement, I dragged him upstairs to my parent's room and as I promised my Papa I would do, I made certain I had his approval this time. Morgan stood at the foot of the bed where my mountain of a Papa sat reading a book. With all the confidence and charm of an expert, Morgan asked for my hand. Papa scowled for a very long minute, looked at me, winked and said, "You got her." It was storybook, fairytale stuff.

Shortly after, I began my Mormon training, called the discussions. Two missionaries, Elder Arvanites and Elder Erickson, came several times a week to "discuss" the doctrines of the church. I was what they called a "golden find," as I had come to many of the deep doctrine conclusions of the LDS church on my own. I was baptized into the faith and thus, became the perfect bride for Morgan. We were young, ambitious, idealistic and both getting what we wanted. It was a flawless arrangement...until it wasn't.

When Morgan's internship ended, the baby and I moved back to our hometown Merced, California into a spare room in his parent's sprawling home. Morgan had turned 19 and that meant it was time for him to leave for his two-year mission. He was sent to Bozeman, Montana, near the place where I was born but had never seen. Elder Erickson continued to phone and make sure I was upholding my LDS status, occasionally, then frequently. Much in the way that Morgan had entered my very tiny circle of trust, slowly and patiently, Elder Erickson soon became my only friend and confidante. I had returned to Merced College to take all the classes I would need to be a good politician's wife, like public speaking and American history and Morgan's parents stepped in as Grands. When Christmas came around, I decided to go back to D.C. for the holidays to visit my parents. It was wonderful to be in a home again that felt like my own and baby Misha and I had our own space in the basement apartment. One evening, Elder Erickson phoned to say that he and his new companion were in the neighborhood and could they bring a Christmas gift for Misha. The doorbell rang, I opened the door and that's when it happened. His crystal blue eyes locked on mine, my stomach did a flip-flop and in a single moment, I fell deeply, madly and hopelessly in love...and so did he. I literally left them standing in that doorway, ran into the kitchen and burst into tears. My fairytale life had just gone up in smoke.

Love is not war. It is not a battlefield. It is neither a task to be accomplished nor a goal to be achieved. Love is permanent. Love is a fact. It cannot be decided upon or left behind. It arrives or it does not. It occupies a space before that space is encroached upon. Love exists. It is already there and it is waiting to be remembered, to be discovered; waiting to be freed. It can be planted like a seed but no matter how that seed is nourished you cannot grow an oak from the seed of a wildflower. My love for Morgan was a wildflower at the foot of that all mighty oak of true-love for Elder Erickson.

I'll fly by the next 13 years of that marriage to the former Elder, Troy Erickson. We stayed in love as the babies came, (four in total) we dreamed dreams and watched plans shatter then ultimately, the catastrophic action of one Mormon Bishop would set into motion the event which destroyed a family.

In the early '90s, still in D.C., I was an executive for a successful sales organization and won several awards for the innovative sales management techniques I devised using a bit of Kenneth Blanchard, a tad of Peter Drucker, some Steve Brown and a touch of Steven Covey. It was a system I used to increase sales closing ratios by up to 30% and earned me a mention in the Who's Who Among Female Executives, 1994. I was on top of my game. I had eight years of activism with Vietnam veterans in various POW organizations and I was hobnobbing with VIPs, like Adrian Cronauer, *Good Morning Vietnam*, "B1 Bob" Dornan and John "Top" Holland. But in the church, I was only a woman. To remain a member in good standing, I was required to attend the Wednesday night lady's homemaking meetings, in which they honed skills like crocheting and cooking. Although I did none of those things in our home, it was required and I went.

One evening, I was walking from one craft room to another with an educated woman I only assumed shared my sense of confinement in the silly meetings and I said, "Well, it took until the '70s for black men to be allowed to hold the priesthood so we can't be far behind." You would have thought I shouted, "Kegger at my house!" A very palpable hush fell over the room and all eyes were on me: Awkward. The next day, the bishop phoned my husband to let him know I was being called before a disciplinary committee and they would see us on Sunday after services. I freaked out. At that time, Adrian had become another dad to me and had seen me through some difficulties. I called him sobbing and, as he always did, he calmed me down with that voice they call second only to James Earl Jones and said, "Is there any evidentiary reason to believe that or is it just your hope?" Hmm. Good question. I spent the next two days pouring over cd rom copies of the Bible, Book of Mormon and the Doctrine and Covenant, their additional scriptures. As it turned out, there was an evidentiary support to my claims. Several of the prophets made reference to one day, women holding the "fullness of the gospel," which in Mormon terms, means priesthood. Adrian helped me to prepare a brief that, when it was finished, was an inch thick. But in that meeting with the bishop, who was a construction contractor, and his Elder's Quorum, they were not about to read it. I sat at that long table with this very angry patriarch breathing fire and I was eight years old all over again, but not as brave. The bishop shouted, "This is *my* flock, as a member of *my* flock you obey your leadership or you face excommunication." I guess it's a good thing I never made it to law school because I busted like a dollar store balloon and sobbed. My husband said, "We'll take it up with the stake president." To sum it up, our little brief went from one leader, up the ladder to the next until it reached the Regional President, J.W. Marriot, who supported the original bishop's ruling. Shut up or get out; face excommunication as an "apostate feminist."

I went home that final day with my life...our life; our spiritual life gasping at my feet. I tried to call Morgan since he was the one who got me into this church, but his sweet wife said he didn't want to talk to me. I don't blame him. But I was lost...so lost, and so was my husband. He wanted to support me more than anything, but he had grown up in the church. It was all he knew.

I walked down to the little creek behind our house and, just as Joseph Smith had done to start this whole mess, I prayed. I prayed from a place I'd never known was even there. Sometime before the sun came up, I had my answer. I am important in the big, big scheme somewhere, somehow, all on my own. That's when the whole card-house of faith I had invested in this one institution collapsed.

Three months later, it was all coming undone. My husband was on a rebellion terror, our family no longer had morning and evening family prayer or Monday family home evenings, and not even my job made sense to me anymore. I picked a fight with the boss's wife that got me fired, so I enrolled in the local community college and Dr. Hal McMullen showed me the light back to a path of spiritual growth that felt like waking up. True to pedagogy, he took us to a Franciscan monastery to chant with the monks, to Tibetan prayer centers, to a Catholic cathedral and a Hindu retreat. He introduced me to Thomas Merton who I came to love so much that I won a regional title as Most Notable Young Philosopher of the Year, in competition with several universities when my paper *Mertonian Mysticism* was published.

I was given the honor to speak at my first east-west dialogue and met two more life teachers who would shape the creature you read before you; a monk and author named Wayne Teasdale and the remarkable nun Ani Kunga. As a matter of fact, a single incident with Ani Kunga in the Tibetan prayer center changed my life paradigm. How could she know that a single conversation which ended in the kitchen that day would remain with me for all the days of my life? Her essence, her wisdom and her spirit are a part of me now. Specifically, the least personal but the most interesting thing I learned was forced on her by my enthusiasm, by which she must have been exhausted. Even at 55, my exuberance when I'm ON to something; a little nugget of knowledge I can feel myself touching with the tips of my fingers; can be annoying. I jump ahead, interrupt and bubble over with, "and then? And then? And then?" So, the conversation ran over until she said, "My lunch is cooking in the kitchen, will you follow me there?" I'm seeing my face on a baby deer bounding behind her. While she spoke in her dulcet and gentle tone, answering whatever tedious question I'd asked with the patience of Buddha, she opened the oven and pulled out...ready? Tibetan nun, remember? A rack of pork chops. I was aghast! Of course, I interrupted, "You eat meat?" As all good adults, she followed my curiosity into this new line of questioning and told the most remarkable story. *"You know, some of the masters teach that to harvest a field of corn is genocide, to walk upon a patch of grass is mass murder. Can it be that an animal has made a karmic pact before entry into a body here, to keep my body alive? I take it upon my karma to consider this. All that we do is by choice. I eat meat by choice."* It totally made sense to me. I could imagine myself in whatever place we come from, assuring my polar bear friend that I would come as a seal to save her pups from starvation. I *so* totally would. And that put to rest the guilt aspect of eating my lovely fellow Earthlings.

I want to add to that lesson all these decades later; I have snakes in my pack. I have a ball python named Argus and a corn snake named Minerva I inherited. When I purchased Argus, I was at a very lonely but workaholic stage of my life and she was the perfect companion. She ate once a week, pooped every 10 days or so and when I walked home from the restaurant/tavern I managed in Canton-Baltimore at 4 am wearing her around my neck, (she was also happy to sleep in my purse for hours at a time), the scariest drug dealers, pimps and street thugs crossed the street to avoid me. She became a true friend with a very distinct and loving personality. But she would only eat live mice. It was another walk of faith for me to sacrifice these little beings for the sake of my friend. I do, however, still and always chant om mani Padme hum, related to the bodhisattva of compassion, as they are facing their fate.

I had truly begun to awaken to a higher state of existence; one that encompassed all walks of life, all stages of development and all species as part of my whole family. But, my husband, the father of my children, the love of my life, came from a very different world. Sadly, my husband didn't deal well with the absence of the comfort and structure of the church, and eventually that house of cards fell down too.

This series of life events, and so many more brought me to you, now, here at this moment with the *whole* of what I have learned. When we meet, and I hope we do; when I take your hand, look in your eyes and engage *your* life's experience, I will inevitably hug you or touch you in such a way that you know I am truly *with* you. This is not put on. It's not a ploy to convert you to my weirdness. It is my reflection of my belief that we are inseparably connected by our Mother, the source of all life, and that makes you mine, as I am yours. I will treat you with honesty and compassion. I will do my damnedest to stay patient and loving and hear your words, even the angry ones with tolerance and humility...until I don't. Because sometimes I won't. I will have bad days and fat days; days and when I feel less than worthy of this shiny human world and I will lose my patience. I'll snap and snarl and spew my potty mouth rant...until I don't. And then, I'll come to you with a cup of tea or a shot of Tullamore Dew and try to begin again. I could apologize for being human but I think apologies mean you won't do it again, and I probably will. So, please...tolerate my human-ness as I endeavor to overcome it and I promise to be the best version of me I can be.

And that brings me to: let's build a bridge of humanity over invisible borders, beyond outmoded tenets and "isms" to save our species and as many others as we can. When we lock arms as one single collective, we can be so much better together...we can.

To paraphrase my beloved Rilke again; when that thing rises up like a shadow in your path, more powerful than you could ever have been in this life and you know it's about to crash down and end your existence as you know it; almost at once, you will see the absolute serenity of nothing, and at that moment, nothing is all that will matter. If we can rise, even to our knees after suffering, we have no choice to look up, out and away and realize we are free.

Chapter 10: Solutions

As much as I would have loved to make this entire book only about solutions (maybe the next one), you had to understand all the intricacies of your contribution to the problems before you would understand the power you hold to fix them. I'm sorry I had to stress you out to get here. But guess what? Hooray, we made it! This is the fun part. We're finished with the bad stuff and we're moving on to all the incredible solutions you brilliant humans have created to solve our biggest problems. Paul and I have come across so many wonderful innovations in our quest to help us fix this mess and below we're sharing some of our favorites. Some of the solutions are ideas, some are intended for institutions and governments but all are meant for you, to give you the ability to change your world. Let's do this.

All my eco-sheroes and heroes agree that the most effective way to save the planet is to **leave half of it alone!** They are supported by the world's science and ecology fronts. At the end of this chapter, we offer our concept of what this will look like. In short, we propose living in small, self-sustaining villages, inside natural terrains or in vertical villages in cities. If you must have a large home, make it an island; completely self-sustaining with a minimal carbon footprint.

Before we get there, you need to know about some of the amazing solutions that are within your reach today. There are a myriad of great ideas but we only had room for our favorites. In order to make them easier to approach, we've broken them down by sector, alphabetized and briefly summarized with links and references. You'll see a lot of crossover; one solution can apply to many of our problems. Some solutions are tiny changes that are easy to find; super affordable but will make massive changes in our impact on the planet. Some are huge solutions that will need your drive and your support to make happen.

We'll tell you how you can help.

Climate Change Solutions:

So, where do we really stand on climate change and what are any of us doing about it? Welp, here is what the U.S. government is doing.

The figure on the next page illustrates the adaptation iterative risk management process. The gray arced circle compares the current status of implementing this process with the status reported by the 3rd National Climate Assessment (NCA) report in 2014. Darker color indicates more activity. Source: adapted from National Research Council, 2010.1 Used with permission from the National Academies Press, ©2010, National Academy of Sciences. Image credits, clockwise from top: National Weather Service; USGS; Armando Rodriguez, Miami-Dade County; Dr. Neil Berg, MARISA; Bill Ingalls, NASA.[263]

How does this techno-speak translate to us plebeians? At the time of the 3rd NCA report in 2014, there was strong awareness of the issues involved in climate change. As their thought process moved toward an in-depth assessment of the situation, focus was quickly diffused by a lack of money or tasks easier to tackle or perhaps just another shiny object. No one knows or at least won't admit the reason.

At the time of the 4th NCA report in 2018, there was strong awareness of the issues involved in climate change. Strong interest was maintained a bit longer into an in-depth assessment of the situation, but when preparing a detailed plan to go forward, their focus was again quickly diffused and morphed into something else. And that folks, is what we call *deja vu*.

---

[263] The National Climate Assessment (NCA) is a United States government interagency ongoing effort on climate change science conducted under the auspices of the Global Change Research Act of 1990. Information taken from their 4th report

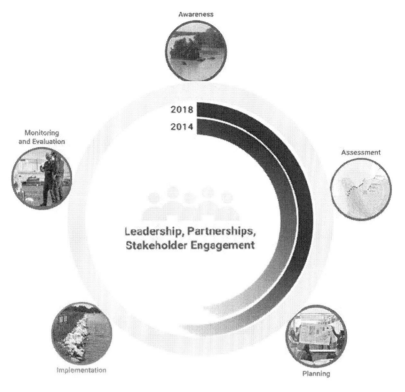

Figure 28 in the 4th NCA Report

As we've seen, climate change is a massively overwhelming concept to wrestle to the ground. But, there are a number of bite-sized chunks we humans can chew off in order to digest its influence on the planet. It starts will research – not reading media posts or watching TV but an extensive examination of the real scientific facts of the situation. There is no middle ground with this fight. You either believe in the truth of science or you are destined to follow the lead of others. Once you have chosen where you stand on this issue, you'll know what you can do.

For those of you who choose to take action, let's take a quick look at how ecosystems work.

Has your car ever broken down on the side of the road? Often, a car breaks down because one simple part is broken. There are many moving parts that make a car work. A spark plug ignites the fuel, wheels move the car forward, and a radiator cools down the engine. In order for a car to work smoothly, all parts need to be intact and in good condition. Just like a car, species living in an ecosystem play an important part in keeping an ecosystem running smoothly. If one species is lost the entire ecosystem can stop working.

Ecosystems are organized in a state of balance where species coexist with other species. If something happens in an ecosystem, it can shift from a state of balance to a state of imbalance. Ecological imbalance is when a natural or human-caused disturbance disrupts the natural balance of an ecosystem. A disturbance is any change that causes a disruption in the balance of an ecosystem.

Examples of natural disturbances are:

- Volcanic eruptions
- Floods
- Natural fires

Examples of human-caused disturbances are:

- The introduction of a new or invasive species
- Logging a forest
- Pollution in all its forms
- Overhunting of a species

Remember that all parts of an ecosystem must work together to exist in harmony. After a disturbance occurs, an ecosystem may recover back to a balanced state. But if an ecosystem has a severe disturbance or is constantly having new disturbances, it may never recover back to a state of ecological balance.[264]

How much of a difference can one person/family really make? Actually,

---

[264] Study.com: What is an Ecological Imbalance?
https://study.com/academy/lesson/what-is-an-ecological-imbalance-definition-lesson-quiz.html

social scientists have found that when one person makes a sustainability-oriented decision, other people do too.

Here are a few examples:

- Patrons at a U.S. cafe who were told that 30% of Americans had started eating less meat were twice as likely to order a meatless lunch.
- In California, households were more likely to install solar panels in neighborhoods that already have them.
- Community organizers trying to get people to install solar panels were 62% more successful in their efforts if they had panels on their house too.

Social scientists believe this occurs because we constantly evaluate what our peers are doing and we adjust our beliefs and actions accordingly. When people see their neighbors taking environmental action, like conserving energy, they infer that people like them also value sustainability and feel more compelled to act.[265]

Every individual counts in saving the planet. No two people have the same emissions. Although the average human releases around 5 tons of CO_2 per year, each country has very different circumstances: developed nations like the U.S. and South Korea have higher national averages (16.5 tons and 11.5 tons per person, respectively) than developing countries like Pakistan and Philippines (around 1 ton each). Even within national borders, richer people have higher emissions than people with less access to goods and services. So, if you choose to take this question into account, you have to remember that …it's where (and who) you are that makes all the difference.[266] Be mindful of who you are and what you can do for planet sustainability.

What can individuals do to create a more sustainable world?

---

[265] BBC – Believe in Humans
http://www.bbc.com/future/story/20181102-what-can-i-do-about-climate-change
[266] BBC – Believe in Humans
http://www.bbc.com/future/story/20181102-what-can-i-do-about-climate-change

Waste less food - Reducing food waste is the number-one thing consumers can do to significantly lessen their climate impact, according to the Project Drawdown report. "Food that is disposed of and spoiled creates methane, and that's why it has an impact on greenhouse gases, because methane is such a strong greenhouse gas," a vice president for Project Drawdown Crystal Chissell says. "And that's why reducing food waste has such a large impact." Food waste occurs when we don't buy produce because it has blemishes or is misshapen, when we discard food because it is a day past the expiration date, or because we simply never get around to eating it, she says.[267]

Obtain your meat from sustainable red meat sources - Factory farms feed cows grains, which cause them to release methane into the air through their gases, says Chissell. "It's not actually natural to their digestive system so it creates more methane," Chissell explains. Chissell says adopting a plant-rich diet, and eating more meat from organic farms where animals are fed natural diets, can help reduce methane. "It's not even necessary to be a vegan or a vegetarian," she says, "it's just reducing the amount of meat that we consume and eating plant-based [foods]."[268] (this is the difference between eating beef and buffalo)

Use less water and energy – This is as easy as turning off the faucet when you brush your teeth, turning the thermostat down / up two degrees seasonally and switching to LED lighting. A little conscious thought goes a long way in conservation.

Change your diet a little – Eat meatless meals a few times a week. Eat organic (pesticide free – most of which are oil byproducts) when you can. Buy local when you can. There are all easy ways to do your part to help the climate.

---

[267] CBS Better - 6 ways ordinary people can prevent climate change, according to researchers and advocates https://www.nbcnews.com/better/science/6-ways-ordinary-people-can-prevent-climate-change-according-researchers-ncna926311

[268] CBS Better - 6 ways ordinary people can prevent climate change, according to researchers and advocates https://www.nbcnews.com/better/science/6-ways-ordinary-people-can-prevent-climate-change-according-researchers-ncna926311

The share of greenhouse gas emissions from animal agriculture is usually pegged at 14.5% to 18%, but the Worldwatch Institute found lots of oversights in those calculations that, when properly counted, bring the ag contribution all the way up to 51%. That, you'll notice, is more than half; which means that after we clean up all the transportation, energy, industry and commerce in the world, we've done less than half the job. The other half is meat and dairy. Refuse to eat it. If this seems too challenging, consider giving it up one day a week. It will still be the most important action you can take.[269]

Reduce and reuse before recycle - Recycling emerged as a virtue before we knew we had a climate problem, and it turns out that transporting and processing materials for recycling is carbon intensive. Recycling still uses less energy than making new products from scratch, but reducing and reusing are even cleaner.[270] Cut out single use plastic conveniences such as straws, single-use water bottles, and eating utensils. If you already have these in your cupboard, reuse them. Wash in warm soapy water and they can be reused many, many more times before proper recycling.

Deforestation

When it comes to the problem of deforestation, the worst offender is agriculture. It goes like this:

- Agricultural (mostly for livestock feed)
- Livestock
- Logging
- Infrastructure expansion (Roads)
- Overpopulation

---

[269] Forbes - 9 Things You Can Do About Climate Change
https://www.forbes.com/sites/jeffmcmahon/2017/01/23/nine-things-you-can-do-about-climate-change/#6644a7a8680c
[270] Forbes - 9 Things You Can Do About Climate Change
https://www.forbes.com/sites/jeffmcmahon/2017/01/23/nine-things-you-can-do-about-climate-change/#6644a7a8680c

As for agriculture and livestock, we'll cover that under "Food." Concerning roads, I'll default back to leaving half the planet alone. Overpopulation? Support Planned Parenthood and wrap that shit up! But logging? It's primarily for wood products and paper so let's talk about logging solutions. Our planet has provided so many resilient and regenerating plants far more sustainable and practical to use than trees. By a long-shot, my favorite is hemp which creates not only delicious whole foods but can be made into a breathable concrete substitute (hempcrete.) And that's not all. Look at what Gaia gave us in hemp:

Hemp

If the Earth provided one solution to all our most gluttonous demands, it would be hemp. This plant has so many uses that farmers in the U.S. could support their families and industry on this single crop. Hemp is such a powerful plant that I don't want to miss anything, so I'm sharing the list directly from experts Ariana Palmieri and Igor Yakovlev:

Hemp Part: Seeds

Hemp seeds can be used for many things. Here's a general list of the most common uses.

- Body care
- Cosmetics
- Soaps
- Balms
- Shampoos
- Lotions
- Foods

Hemp Seed Hearts

- EFA food supplements
- Hemp protein powder
- Hemp seed oil

- Industrial products
    - Coatings
    - Oil paints
    - Solvents
    - Varnishes
    - Fuel
    - Printing inks
    - Nail polish

Hemp Part: Stalk

The stalk of the hemp plant is where our Cannabidiol (CBD) oil comes from. Currently fourteen states within the U.S. can legally grow and process industrial hemp and hemp seeds. These states include California, Colorado, Kentucky, Oregon, and Tennessee. CBD oil is grown organically in Colorado where it's extracted from the stalks and stems of industrial hemp plants with 0.3% THC or less. It is completely safe to use and will not create a high. Here are a few other ways hemp stalks can be used (besides CBD oil):.

- Paper
- Packaging
- Printing
- Cardboard
- Newsprint
- Textiles
- Fine fabrics
- Clothing
- Shoes
- Diapers
- Denim
- Handbags
- Building materials (hempcrete)
- Fiberglass substitute

- Fiberboard
- Insulation
- Acrylics
- Industrial textiles
- Molded parts
- Rope
- Caulking
- Canvas
- Netting
- Tarps
- Carpeting

Hemp Part: Leaves

Believe it or not, the leaves of the hemp plant can be used too, although they aren't as versatile as the other parts. They are very absorbent, which makes them good for animal bedding and great for adding to compost and mulch.

Hemp Part: Roots

The roots of the hemp plant don't go to waste either, they can be used for:

- Joint pain
- Arthritis
- Eczema
- Fibromyalgia
- Organic compost and nutrients

There are so many uses for hemp I'm not providing a single link for purchase or reference.

Bamboo

Another incredibly versatile and sustainable alternative to trees is bamboo. It can be used to create:

- Charcoal
- Alcohol
- Bed sheets
- Blinds
- Paint brushes
- Bicycles
- Cutting boards
- Clothing
- Fabrics
- Flooring
- Garden plants
- Matting
- Instruments
- Record player needles
- Roofing
- Umbrellas
- Wedding favors
- Sugar (as in sugar cane)
- Deodorizers
- Beer
- Beehives

A great place to start on your quest to build a bamboo world around you is:

https://www.bamboogrove.com/bamboo-products.html

Recycled Paper

This is straight from my favorite resource on recycled paper, Green America's Better Paper Project:

Recycled paper is proven to use less tree fiber, fresh water, energy, and produces less waste than traditional virgin fiber paper. One ton of magazine paper made from virgin fiber requires fifteen trees. Those trees can remain in the ground if we widely and effectively increase our use of recycled paper. The Better Paper Project's overall goal is to see a reduction in our consumption of paper products. However, we promote the use of "better paper" when necessary, meaning high recycled content and certified by the Forest Stewardship Council.

The Better Paper Project is also proud to represent Green America on the Environmental Paper Network (EPN), an alliance of organizations working to address challenges and opportunities for social justice and conservation presented by the expanding forest, pulp, and paper industry.

You can find out more and support their efforts here:

https://www.greenamerica.org/save-trees

Wood Composites from Used Plastic

One of our best hopes to minimize waste is to use what we already have; called Circular Economy. Wood-composites combine the look of wood with the functionality and manufacturing ease of plastics. These wood-plastic composites blend fine wood particles with renewable, biodegradable, reclaimed, recycled, or virgin plastic materials to make firm, smooth pellets for convenient handling and further processing. Along that line is combining wood with use plastic. This is called a wood composite and this material can be used to create boards, bricks and planks which are weather resistant. You can learn more at:

https://www.greendotbioplastics.com/materials/wood-composites/

Recycled Plastic

Large-scale construction projects can be major sources of waste or recycling, depending on the materials. Better building material selections can benefit the environment without sacrificing integrity. For example, plastic lumber looks and installs like traditional lumber, but is primarily sourced from post-consumer and post-industrial products. This engineering breakthrough came out of Rutgers University with the goal to recycle the entire world's unused plastic. Thomas Nosker figured out a way to harness the power of non-toxic synthetic plastics to create a new building material. The invention of plastic lumber is made mostly from household milk jugs and laundry detergent bottles. Recycled plastic lumber is economical, offers the same look as wood, as well as:

- Weather-Proof
- Durable
- Low-Maintenance
- Sustainable
- Vandal-Proof

You can learn more and purchase this product at:

https://plasticboards.com/green-solution-for-building-materials/

Cork

Cork is the outer bark of the cork oak tree, *quercus suber*, which grows mainly in the Mediterranean region. Cork oak is a protected species. The bark is a vegetal tissue composed of an agglomeration of cells filled with a gaseous mixture similar to air and lined with alternating layers of cellulose and suberin. In a context of increasing concern for the environment, cork remains the only tree whose bark can regenerate itself after each harvest — leaving the tree unharmed. It is truly a renewable, environment.[271] It is non-toxic and does not off-gas VOCs. It gives a warm, cozy and organic feel to an interior space. Each piece of bark is unique in texture, grain and color, for a one of a kind look. Cork satisfied Leadership in Energy and Environmental Design (LEED) credits for being renewable, recycled and having low emissions.

Because corks as bottle stoppers are being replaced with plastic corks or screw tops, the cork industry is slowing down. You can help keep this sustainable industry alive and keep thousands of people working by

---

[271] Green Building Supply: Cork 101 https://www.greenbuildingsupply.com/Learning-Center/Flooring-Cork-LC/Cork-101

buying cork products. There are many uses for it around the home, and now furniture designers are playing with it, since it's so lightweight.[272]

Cork's elasticity, combined with its near-impermeability, makes it the perfect material for making:

- Flooring (has a wonderful bit of give)
- Rigid but breathable insulation (fire resistant)
- Exterior finish (and resists mold and mildew)
- Floor underlayment (nice and cushy)
- Acoustic wall coverings and countertops (it has antimicrobial properties)
- Bottle stoppers
- Bulletin boards

Learn more and find products here:

https://www.greenbuildingsupply.com/Learning-Center/Flooring-Cork-LC/Cork-101

https://www.builddirect.com/blog/is-cork-the-most-sustainable-building-material/

Soy Plywood

As solid wood is expensive and often has inconsistent cross-sections, plywood and fiberboards are frequently used for interiors in construction. But the adhesive in plywood is made from formaldehyde which is unhealthy not only for the environment and our general health; it's dangerous for the factory workers. Luckily, there is a soy-based adhesive to make formaldehyde free plywood.

Learn more and find products here:

https://www.columbiaforestproducts.com/product/purebond-classic-core/

---

[272] Build Direct: Is Cork the Most Sustainable Building Material?
https://www.builddirect.com/blog/is-cork-the-most-sustainable-building-material/

Nutshells

Walnut shells can be used for cleaning and polishing, as filler in dynamite, and as a paint thickening agent. Shells from pecans, almonds, brazil nuts, acorns, and most other nuts are useful in composting. Their high porosity makes them also ideal in the production of activated carbon by pyrolysis (decomposition by heat). Shells can also be used as loose-fill packing material, to protect fragile items in shipping.

Black Walnuts are very exciting as they solve many problems we've already addressed regarding wastewater and fossil fuel hazards. Check this out:

Abrasive Cleaning

When ground, Black Walnut shell can be used as abrasive for a variety of cleaning and polishing applications. And because of its breakdown resistance, it can be reused many times resulting in reduced cleaning costs.

Water Filtration

Black Walnut shell serves as an environmentally safe and effective filtration media for separating crude oil from water.

Lost Circulation

The oil drilling industry uses Black Walnut shell as a key ingredient in making and maintaining seals in fracture zones and unconsolidated formations.

Please visit Hammons Black Walnuts for more information:

https://black-walnuts.com/

Straw

After a grain harvest, a huge amount of stalk fibers are wasted. Wheat straw can be used as fibrous filler combined with natural polymers to produce sustainable composite lumber as well as enviroboard, a sustainable packing and a heavy canvas like fabric. I didn't find a single source but poke around out there for what may fill your needs in this line of alternatives.

Energy

Introduction by Paul Hollis

Energy is the heartbeat of civilization. There can be no argument on that point. Since the invention of the power grid in the 19th century, it has not only maintained human life but pretty much controlled it, too. Ask anyone when the power goes out in an ice storm, when cable TV is having a power induced seizure, or when the cell signal is blinking "no signal" and they're about to begin negotiations with the devil.

Pamela and I recently had an informative conversation with an energy expert with more than 50 years in the energy business. Rocky Buldo, of https://enoughtechnologies.org, is dedicating his genius to demonstrating how to power individuals to be energy independent. Here is a little of what he taught us.

There are two energy domains: Alternating Current (AC) and Direct Current (DC), each with very different properties. AC is an artificial process that creates a flow of electrons which reverses its direction of flow at regular intervals in a conductor while DC is naturally flows in one direction – to a device requiring power to operate.

The United States created the centralized power grid in the 19th century as our perceived best option. Buldo says, "Over time, however, the North American power grid has become the largest, most complicated, and intensely fragile mechanism ever created by man." It is expensive to keep it running, impossible sometimes to maintain and as we've seen, unreliable at the most inopportune times. The grid is becoming highly susceptible to outside forces, not the least of which is weather aberrations and foreign intervention.

European power companies were quick to identify inherent issues in an alternating current system and gravitated to 230V direct current. Perhaps the biggest flaw surrounding alternating current is that its energy cannot be stored unless it is converted DC or heat. Consequently at the end of the day, literally, AC current must be burned off.

The reason for this short horror story is to basically tell you there are solutions to *business as usual* power. Enter Rocky Buldo. He is not, nor are we, advocating dissolving or discontinuing the power grid. This is a process of evolution, not revolution, and major change will come slowly.

As we prepare to take our first small steps away from grid power, Rocky wants us to know ordinary consumers cannot own and store alternating current power. The government is all over that and any number of laws that control its distribution, handling and use. The good news is that, for now, direct current can be owned and stored by us ordinary citizens.

We have focused a great deal on the circular economy in this publication, and energy is a part of that base concept. Energy moves in a direct path in nature, a Direct Current. We can mimic nature in our energy practices just like a tree that receives energy directly from the sun and converts it to life. If we can change what we do, bit by bit, we can become less burdensome and more independent of the grid. After all, the opposite of dependence is freedom. What you should have is the power to choose which source of energy you use.

We learned a lot about the need for a smart power grid and how it may be rebuilt sometime in a future generation. But for now, we will discuss a number of practical energy solutions available to us today.

Solar

We currently believe solar energy is the best alternative to replace fossil fuel as the major energy source for a few simple reasons; 1) solar power is renewable at absolutely no cost and it can supply energy indefinitely, and 2) the use of solar panels is more environmentally friendly than some other options we'll discuss here.

Solar panels absorbing sunlight with photovoltaic cells, generating direct current energy and converting it to usable alternating current energy with the help of inverter technology. Alternating current energy then flows through the home's electrical panel and is distributed accordingly. Here are the main steps for how solar panels work for your home:

- Photovoltaic cells absorb the sun's energy and convert it to DC electricity that is stored in batteries
- A solar inverter converts DC electricity from your solar modules to AC electricity, which is used by most home appliances
- Electricity flows through your home, powering electronic devices
- Excess electricity produced by solar panels is fed to the electric grid.[273]

Our favorite solar solution is the circular set of panels that can track the sun's movement across the sky. Here is one such solution called the Smartflower.

The Smartflower's name comes from its design – the solar cells are arranged on individual "petals" that open at the beginning of each day and look like a flower (hence, "solar flower"). After the sun goes down, the Smartflower's petals fold up and a self-cleaning process kicks in.

---

[273] Energy Sage: How do solar panels work? https://news.energysage.com/solar-panels-work/

In addition to solar cells, the Smartflower system contains a dual-axis tracker that makes it possible for its petals to follow the sun across the sky throughout the day. Thanks to this tracking capability, the Smartflower can produce significantly more electricity than a similarly-sized rooftop solar panel system – up to 40% more, according to Smartflower's website. The 12-petal, 194 square-foot structure comes with 2.5 kilowatts (kW) of electricity production, which is roughly equivalent to a 3.5 to 4 kW fixed rooftop array.

While there is currently just one solar flower version available in the U.S., the Austria-based company has plans for two additional models with new features. The Smartflower PLUS will have the same solar electricity generation benefits of the standard Smartflower, plus an integrated battery that provides 6 to 13 kWh of energy storage. It was expected to be available to U.S. customers in early 2018, but the actual rollout has been unclear. There are also plans to bring a Smartflower with integrated electric car charging to the U.S. market (launch date still to be determined).[274]

For more information, please visit their website at https://smartflower.com.

Hydrogen Energy

Through the magnificent process of photosynthesis, life as we know it is fueled by energy from the sun. The release of oxygen into the atmosphere by photosynthesis forms biomass (organic matter used as fuel) from the water and CO_2 which enters the leaves of plants. Within the circular economy of nature, when living organisms require fuel, they rejoin oxygen to the biomass to re-form CO_2 and water. We call this metabolism. Hydrogen gas has the largest energy content of any fuel,

---

[274] Energy Sage: Smartflower solar: the complete solar flower review
https://news.energysage.com/smartflower-solar-complete-review/

Pioneer Energy is a Colorado based service provider and original equipment manufacturer that solves gas processing challenges on the oilfield through development and deployment of new technology.

The company offers a range of standard gas capture and processing units for monetizing tank vapors, associated gas, and non-associated gas, and for conditioning gas streams. Fractionators, knockouts and separators are also in our product portfolio. The company also specializes in providing custom engineered solutions at any desired scale to accommodate customers' unique field gas conditions, while providing high-quality output streams that meet stringent requirements.

Pioneer Energy's Flarecatcher™ line of units provides associated and non-associated gas capture and processing at the well site with superior performance than any other modular system, producing NGLs, pipeline quality lean methane, and enabling producers to achieve regulatory compliance.

Pioneer Energy's Vaporcatcher™ and Dew Point Suppression line of units capture tank battery vapors and extract NGLs at high yield instead of sending these valuable commodities to flare. This dramatically cuts emissions while simultaneously providing significant economic return. These systems can also be used in gas conditioning applications, cooling associated gas and removing water and heavy hydrocarbons prior to injection into a gathering system. Condensables can also be removed by the system in fuel gas conditioning applications.

Pioneer Energy's engineering, field service, and remote operations teams provide best-in-class support of both domestic as well as international customers' operations. The company's knowledge is underscored by their field experience in the Bakken and Denver-Julesburg oil fields.

The company continues to build on our knowledge, refining our systems while developing new solutions to meet tomorrow's gas processing needs for the oil and gas industry.

For more information, join us at (www.pioneerenergy.com).

making it a very good 'vehicle' for holding and distributing that energy. By using electricity, we can mimic this process to split water into hydrogen and oxygen. The oxygen is released into the atmosphere and the hydrogen can now be stored. To use this as a fuel source, we need to rejoin the H2 with oxygen (O) in a *fuel cell*. In the cell, H2 electrons arrive at a cathode where they are recombined with hydrogen in the presence of the reactive gas oxygen which provides the energy to form water again. The electrodes can be made of porous carbon nanofibers coated with a catalyst such as cobalt or nickel[275] which enables the cell to covert almost all of the chemical energy to electricity.

Hydrogen is the most abundant element in the universe, but here on Earth it's a bit of a cling-on; only found in the company of another element(s). Although it is a powerful source of energy, the task of separating it from its companions is difficult and expensive, much like a Hollywood divorce. But recent advances which enable us to use that cobalt for a catalyst instead of platinum have made it within our grasp.[276]

When it comes to global energy, we're in a heap of trouble. For the U.S., we are stuck with aging infrastructures with fragile and daunting maintenance challenges just to keep it running. Those patch-work fixes add to the rising costs of utilities for the end user. But on almost every solution-oriented front, chemical processes are leading the crusade. There's a new kid in town and Bloom Energy's servers convert fuel into electricity through an electrochemical process without combustion.

Bloom boxes are modular building blocks that work in clusters of various configurations based on need. Lots of boxes = lots of energy. These solid oxide fuel cells consist of three parts:

1. Electrolyte
2. Anode electrode

---

[275] Phys.org: New material creates fuel cell catalysts at a hundredth of the costhttps://phys.org/news/2018-01-material-fuel-cell-catalysts-hundredth.html
[276] Phys.org: New material creates fuel cell catalysts at a hundredth of the costhttps://phys.org/news/2018-01-material-fuel-cell-catalysts-hundredth.html

3. Cathode electrode

In a fuel cell, an electrolyte, a solid ceramic material, is what the ions move through. The anode and cathode are made from an ink that coats the ceramic electrolyte. There are no precious metals, corrosive acids or molten materials. An electrochemical reaction converts fuel and air into electricity without combustions. The system can use natural gas, hydrogen or a combination of the two to operate. As fans of hydrogen technologies, we were happy to see that in June 2019, Bloom Energy announced Hydrogen-Powered Energy Servers to Make Always-On Renewable Electricity a Reality.[277]

You can learn more at:

https://www.bloomenergy.com/

An excellent source for all things hydrogen, including generators, is the *Fuel Cell Store*:

https://www.fuelcellstore.com/

HYDROGEN CARS

Much in the way we have seen the fossil fuel moguls fighting like wildcats to damage the reputation of renewable energy for decades; the auto industry has been fighting the change to renewable cars. Besides being major contributors to CO_2 emissions, everything our cars produce right now from that slow drip leak of coolant to the particulate matter released in the air; is damaging. We need to fix what we can as quickly as we can. Electric and hydrogen cars are a great start for many reasons. The most efficient way we can hang on to our cars is to have them run on hydrogen, oxygen and sunlight. A car that runs on hydrogen can drive 500 miles before it needs recharging.

---

[277] Bloom Energy: Bloom Energy Announces Hydrogen-Powered Energy Servers to Make Always-On Renewable Electricity a Reality
https://www.bloomenergy.com/newsroom/press-releases/bloom-energy-announces-hydrogen-powered-energy-servers-make-always-renewable

Please visit our friend Kevin Kantola (HydroKevin) to learn more about hydrogen cars and to see the industry giants now on-board:

http://www.hydrogencarsnow.com/index.php/about/

Wind Power

There are three main types of wind energy:

1. Utility-scale wind: Wind turbines that range in size from 100 kilowatts to several megawatts, where the electricity is delivered to the power grid and distributed to the end user by electric utilities or power system operators.
2. Distributed or "small" wind: Single small wind turbines below 100 kilowatts that are used to directly power a home, farm or small business and are not connected to the grid.
3. Offshore wind: Wind turbines that are erected in large bodies of water, usually on the continental shelf. Offshore wind turbines are larger than land-based turbines and can generate more power.

For the first option above, standing at least 80 meters (262 feet) tall, tubular steel towers support a hub with three attached blades and a "nacelle", which houses the shaft, gearbox, generator, and controls. Wind measurements are collected, which direct the turbine to rotate and face the strongest wind, and the angle or "pitch" of its blades is optimized to capture energy.

This type of turbine will start to generate electricity when wind speeds reach six to nine miles per hour and will shut down when the wind blows too hard (55 mph). Over the course of a year it can generate usable amounts of electricity over 90% of the time.[278]

However, we cannot in good conscience endorse it because of two major concerns:

---

[278] American Wind Energy Association: What is wind energy?
https://www.awea.org/wind-101/basics-of-wind-energy

1. One study found that 25% of wind turbine faults caused 95% of the downtime. Reliability of wind turbines has improved with time and has achieved an availability of 98%, but wind turbines fail at least once per year, on average, with larger wind turbines failing relatively more frequently.[279]
2. Canadian family physicians can expect to see increasing numbers of rural patients reporting adverse effects from exposure to industrial wind turbines (IWTs). People who live or work in close proximity to IWTs have experienced symptoms that include decreased quality of life, annoyance, stress, sleep disturbance, headache, anxiety, depression, and cognitive dysfunction. Some have also felt anger, grief, or a sense of injustice. Suggested causes of symptoms include a combination of wind turbine noise, infrasound, dirty electricity, ground current, and shadow flicker.[280]

In addition to these listed health issues, we have also heard unconfirmed reports from a prominent American politician that wind turbines also cause cancer. While we appreciate his concern, let's be clear, there is absolutely no evidence to support this conspiracy theory.

The second option, distributed or "small" wind turbines, seems like it would make a decent solution for a small community because distributed wind turbines are not connected to the grid. As a result, energy usage estimates can be anticipated and planed for best community use.

We just came across *New World Wind*, a progressive wind-energy-based company, with a unique set of wind power offerings in the distributed power market. Gaining inspiration from nature to produce tomorrow's sustainable energy, New World Wind offers solutions that combine high technology with smart design, micro wind turbines with solar petals. A range of innovative products, elegant and completely silent, makes a

---

[279] Exponent: Wind Turbine Reliability
https://www.exponent.com/knowledge/alerts/2017/06/wind-turbine-reliability
[280] Havas M, Colling D. Wind turbines make waves: why some residents near wind turbines become ill. Bull Sci Technol Soc. 2011;31(5):414–26.

seamless fit in both urban and natural environments.[281]

Its patented Wind Trees, combining wind and solar technology, can be "planted" almost any place where beauty is desired over the often roughness of stark technology. Colorfully delicate leaves are mounted on metal branches and trunks; Wind Trees blend into both natural and urban landscapes and can generate power to fit small community needs.

The third example, offshore wind, turns out to be the original concept of an individual wind turbine on steroids; an entire field (or farm) of wind turbines floating on the flat surface of deep water with nothing to inhibit the wind and thus there is much greater electrical output. Obviously, because option one and three above are both inevitably connected to the grid, they do not fit into our definition of a community power source.

Wave Power

Ocean waves have a lot of energy, and wave power is a renewable energy source capable of generating electricity. Wave power is produced by the up and down motion of floating devices placed on the surface of the ocean. Wave power is free, sustainable, renewable, and produces zero waste. Therefore, it can contribute to reducing our carbon footprint. There is an enormous energy potential that can be taken out of the waves and tides, but unfortunately at our publication date scientists, companies, and national authorities have not yet understood how to make it a viable product.[282]

Until feasible solutions come along and water levels increase so that we all can have ocean front property, we will leave this solution to future discussions.

---

[281] New World Wind: http://newworldwind.com/en/wind-tree/
[282] Surfer Today: How does wave energy work?
https://www.surfertoday.com/environment/how-does-wave-energy-work

Hemp Batteries

When it comes to energy, the most important aspect for long-term is the ability to store it. Cavemen figured this out when they realized rocks retained heat from a fire. Our present conventional batteries come with a lot of downfalls, like exploding corrosive contaminants leached into soil when they are discarded, and generally not doing the very thing they were designed to do, efficiently. In almost every solution we have offered, you have seen mention of hemp. This single crop could save our farmers. Well, even in the arena of energy storage, this amazing plant does it better.

In 2014, scientists in the U.S. found that waste fibers (shiv) from hemp crops can be made into "ultrafast" supercapacitors that are "better than graphene"; a synthetic carbon material lighter than foil yet bulletproof. Graphene is a great material but its exorbitant cost makes it prohibitive while hemp fibers come in at one-thousandth of the price.[283]

A fascinating article on EARTHTECHLING tells us that in February 2019, Alternet Systems hired David Mitlin, a professor at New York's Clarkson University who has been researching hemp for energy storage for years. Mitlin's research uses hemp bast, the inner bark of the hemp plant and a waste product during hemp production, as a replacement for graphene. We Endangered Earthlings are cheering on the sidelines, hoping they bring this technology into the hands of we tech-addicted humans.

You can read the whole article here:

https://Earthtechling.com/energy-storage-supercapacitor-hemp-waste/

Food

In *The Virus Human,* we discussed the fact that trying to feed 10 billion

---

[283] BBC News: Hemp fibres 'better than graphene 'https://www.bbc.com/news/science-environment-28770876

humans using outmoded farming methods is mathematically impossible; we simply don't have the resources. But this is the place we get to discuss all the great innovations that will make it possible. There are other methods of farming on the rise, but we're highlighting our favorites below.

In January 2019, a report published by the Ellen MacArthur Foundation at the World Economic Forum Annual Meeting in Davos tells us that 80% of all food is expected to be consumed in cities by 2050. If you live in a big city, you probably don't see a lot of farming going on which makes it hard to shop locally grown food. The methods below make vertical farming an attainable future. We're not providing links, as you will find many options out there to support your individual needs.

Hydroponics

Hydroponics is a farming method that uses up to 95% less water. The plants are not grown in soil, they are essentially dangled above nutrient rich circulating water which allow the roots direct access to their food source. This makes it possible to adapt the nutrients to the individual needs of the plant; something not easily done in soil. There is no irrigation runoff and dispersion from fields because the water stays right there with the plant where it belongs. There is also no exposure to soil-bound diseases and pests and weeds, which means fewer chemicals. Just to mess you up with a nasty visual, there are no farm workers pooping and whizzing in the same dirt as your food. E. coli anyone?

This method makes it possible to create vertical crops anywhere in the world, even the desert. In addition, the roots don't have to expand outward in search of nutrients, so they can use their energy to grow up, providing a harvest with up to eight times the yield of traditional farming. This also enables the plants to be placed closer together so they can be grown in an indoor garden small enough to fit in the corner of an apartment or a crop tall enough to fill a skyscraper. Every home in the world is capable of providing food for its inhabitants. This is especially crucial in cities full of food deserts, where access to groceries and produce is limited, at best.

Aquaponics

By using both plants and fish in a symbiotic relationship inside a closed loop system, we can grow the whole food chain in one place. Basically, the fish doodles provide nutrients and good bacteria for plants in a hydroponic grow unit. In turn, he plants purify and filter the wastewater that gets recycled directly to the fish ponds. It's a great solution for sustainable fish farming and one of our favorite solutions for average citizens alike.

LED Hydroponics

This is really exciting. In Japan, Shigeharu Shimamura turned a former Sony semiconductor factory into the world's biggest indoor hydroponic farm. Shimamura's yields are 100 times more fruitful than conventional farming methods. The secret is in the LED lighting. Different colors of light travel at different speeds or wavelengths. When it comes to plants, what's good for the seedling isn't necessarily good for the adult. So using different colored lights for different plant species and stages of growth, Shinanmura can actually accelerate plant growth and yield tremendously.

Nearly 17,500 LED lights over across 18 cultivation racks that are 16 levels high are used in conjunction with regulated temperature and humidity levels within this productive grow room. The result is that 25,000 of square feet produce 10,000 heads of lettuce per day. By controlling the wavelengths of light, the cycles of days and nights have been shortened to accelerate growth rates. He also uses a core-less lettuce variant to reduce waste.

You can learn more about this amazing technology in this article with multiple links and references:

https://weburbanist.com/2015/01/11/worlds-largest-indoor-farm-is-100-times-more-productive/

Urban Eco Villages

When you are finally ready to recognize the mathematical certainty that there are not enough resources on planet Earth to support 10 billion humans at our present rate of consumption, then you are ready to consider solutions. We happen to agree with most of the world's ecologists who say that the way to save Earth is to leave half of her alone. So, where does that fit in to the certainty that we will run out of space before humans? The only solution is to build up. It is also our collective opinion that we should live within nature in small villages. Soon you'll come to our concept of a vertical village, but as far as food goes, this solution, *Plantscrapers*, is right up our alley. Our concept is shared by many innovative organizations around the world but in essence, an urban ecovillage would provide housing, office space, retail space, vertical gardens using aquaponics systems for protein and renewable energy sources within shared community standards in a vertical environment.

You can learn more about these innovative approaches to urban living at:

https://ecovillage.org/urban-ecovillages-north-america/

Cellular Agriculture (#CleanMeatBaby!)

Hooray! It's time. One of our favorite global solutions to so many of our problems is #CleanMeatBaby! The name "clean meat" is taking all kinds of flack, so to be clear, we are talking about Cellular Agriculture or Cell-Ag for short. The demand for meat is projected to rise 70% by 2050, and this demand cannot be met in a sustainable manner by current livestock farming methods. But the science or practice of farming animal products from cells rather than entire animals, providing the factory is using sustainable energy, is the way forward for carnivores. Cellular-Ag can produce, but is not limited to, food animal products like meat, milk, and eggs, as well as leather, silk and rhinoceros horn.[284] Cellular agriculture

---

[284] CAS: Cellular Ag 101 https://www.cellag.org/cellag101/

focuses on the production of agriculture products from cell cultures using a combination of biotechnology, tissue engineering, molecular biology, and synthetic biology to create and design new methods of producing proteins, fats, and tissues that would otherwise come from traditional agriculture.[285] In short, you can grow meat in a vat to feed millions. And keep in mind, we don't have to use a huge amount of food that only creates a small amount of food. Each year an estimated 41 million tons of plant protein is fed to U.S. livestock to produce an estimated 7 million tons of animal protein for human consumption. About 26 million tons of the livestock feed comes from grains and 15 million tons from forage crops.[286] By 2050, this cycle is no longer viable.

Via cellular-Ag, cells are given media in which to grow, a shape to fill and viola, meat. It is the same cells as that steak you wanted in the first place. This method is cruelty free, disease free and far more sanitary. And remember the scary truth we discovered about the leather industry? Even leather can be made through cellular-Ag without killing the animals.

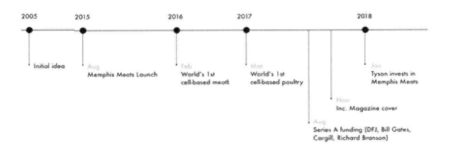

You may have already read our gray box profiling *Memphis Meats,* but to give you an idea of the viability this solution, take a look at this impressive timeline:

---

[285] Wikipedia: Cellular agriculture https://en.wikipedia.org/wiki/Cellular_agriculture
[286] Cornel Chronicle: U.S. could feed 800 million people with grain that livestock eat, Cornell ecologist advises animal scientists http://news.cornell.edu/stories/1997/08/us-could-feed-800-million-people-grain-livestock-eat

Cellular-Ag is already commercially available in some places and the big boys like Bill Gates, Richard Branson and Tyson Foods have already jumped on the bandwagon, but to learn more about the process, products and potential investment opportunities, please visit:

https://www.CellAg.org/

Vegetable-Based Meat

By Pamela Dawn

Okay yeah, so technically it's not meat, but as a meat eater I can tell you it's delicious and satisfies my burger craving. There are several out there but below are my two favorites.

Beyond Meat

Protein, fat, minerals, carbohydrates, and water are the five building blocks of meat. Beyond Meat sources these building blocks directly from plants, (pea protein isolate,) to create delicious, mouthwatering plant-based meat that are so meat-like they bleed when you cook them.

They have delicious products like burger, crumbles and sausage to create a variety of dishes. Mmmmm tacos!

These products are available in more than 58,000 grocery stores, restaurants, hotels, universities. You can visit their site to find locations near you:

https://www.beyondmeat.com/

Impossible Foods

The first time I had the Impossible Whopper at Burger King it was a religious experience. There were tears of joy there was drooling and lots of juice running down my arm. It's so divine.

Impossible Foods say they discovered what makes meat taste like meat then they figured out how to make meat from plants. Impossible Burger is their first product, made from plants, (heme/gmo, soy and potato,) for people who love ground beef with the same flavor, aroma and nutrition we know and love.

Besides Burger King, you can find the Impossible Burger at 17,000 locations. Check the site below to find one near you:

https://ImpossibleFoods.com

Medicine

This book is about what we average citizens can do right now to change our world. When it comes to medicine in the U.S., our best and greatest can't seem to come to a viable solution. So we Endangered Earthlings won't begin to offer a blanket solution. What we will do, as we have tried to do with this entire publication, is to give you the power to help yourselves.

When it comes to our bodies, most of us know them pretty well, having lived in them since we came to Earth. There are many natural methods to approach minor health issues, but the medical institutions are always the place to go for testing, diagnoses and long-term treatment. We can, however, fine tune the functioning of our bodies and keep a cabinet of natural remedies on hand for the small stuff. KEEP IN MIND: many herbal and natural substances react adversely to many medications and vice versa. You absolutely must consult your physician before attempting to treat with herbs, supplements and alternative medicine.

That having been said, here are some very cool ways to fine tune the organism that is you.

This website is a wealth of health topics and does a great job filling in some of the alternative options in the field of health. Go to:

https://www.urmc.rochester.edu/encyclopedia.aspx

Prevention published an awesome article on 25 healing herbs you can use every day. We can't stress enough, check with you doctor first.

https://www.prevention.com/life/a20438272/25-healing-herbs-you-can-use-every-day/

One of our favorite sites to use as a reference when it comes to survival plants is a wonderful website called EatTheWeeds.com. Pamela often shares dishes and decoctions on her social media from things she samples based on information from this weed guru.

http://www.eattheweeds.com/

We already made mention of the remarkable uses from the hemp plant. For centuries, marijuana and hemp plants have been used for healing. To clarify the difference between marijuana and hemp, while they share the same taxonomy, "hemp" generally refers to non-intoxicating cannabis that is harvested for the industrial use of its derived products. These are a couple of awesome sites to learn more:

WEBMD, A TO Z GUIDE TO MEDICAL MARIJUANA

https://www.webmd.com/a-to-z-guides/medical-marijuana-faq

7 BENEFITS AND USES OF CBD OIL (PLUS SIDE EFFECTS)

https://www.healthline.com/nutrition/cbd-oil-benefits

As a bit of additional background in this booming industry, Cannabidiol (CBD) oil is a product that's derived from cannabis. It's a type of cannabinoid, which are the chemicals naturally found in marijuana plants. Even though it comes from marijuana plants, CBD doesn't create a "high" effect or any form of intoxication — that's caused by another cannabinoid, known as THC.

There's some controversy around cannabis products like CBD oil because of recreational marijuana use. But there's growing awareness about the possible health benefits of CBD oil. Here's what you need to know about six potential medical uses of CBD and where the research stands[287]:

- Anxiety relief - CBD may be able to help you manage anxiety
- Anti-seizure - CBD has been in the news before, as a possible treatment for epilepsy
- Neuroprotective - Researchers are looking at a receptor located in the brain to learn about the ways that CBD could help people with neurodegenerative disorders
- Pain relief - The effects of CBD oil on your brain's receptors may also help you manage pain
- Anti-acne - The effects of CBD on receptors in the immune system may help reduce overall inflammation in the body
- Cancer treatment - Some studies have investigated the role of CBD in preventing cancer cell growth, but research is still in its early stages.

Though beneficial results of CBD oil are now surfacing, research studies are still slow to catch up. This is largely due to back-seat driving of big pharma; you know, the people who fund such studies. As qualified findings confess to natural remedies, the less chemically produced drugs are worth. Until more studies are available, we suggest you buy a two-seater, eliminating the backseat, and do your own research. We will leave you with this one downside in today's current marketplace:

CBD is being produced without any regulation, resulting in products that vary widely in quality, said Marcel Bonn-Miller, an adjunct assistant professor of psychology in psychiatry at Pennsylvania University, School of Medicine.[288]

In the end, when nothing else helps alleviate your pain, you will either try it or not; your decision. We believe natural remedies are far better

---

[287] Healthline: 6 Benefits of CBD Oil
https://www.healthline.com/health/cbd-oil-benefits
[288] WebMD: CBD Oil: All the Rage, But Is It Safe & Effective?
https://www.webmd.com/pain-management/news/20180507/cbd-oil-all-the-rage-but-is-it-safe-effective#1

than chemical mixtures with lists of side effects as large as its price.

Sound Therapy

Have you ever been in a really sour mood then had your favorite song come on? Music is a natural healer. Essentially, our bodies are electro-chemical organisms. Beyond just the sounds of music, specific vibrations and frequencies are believed by many to be a very effective healing aid.

You can learn more about this fascinating healing art here:

http://sciencenotes.ucsc.edu/2015/pages/sound/sound.html

Ocean Pollution

We covered the bulk of our waste issues in *Waste Management* and *Plastic* and *Wastewater* below so here we will focus on some of the other forms of ocean pollution.

Acoustic Pollution

To truly understand the magnitude of this problem, we recommend you watch the documentary *Sonic Sea*. As we previously mentioned, many species of life depend on sound in the ocean to survive and the sounds we humans make in the sea are not only disruptive to life cycles; for some, it is deadly.

But we have within our reach the technology to change this. Air guns are standard equipment for seismic mapping used frequently by the oil and gas industry, and they are the second worst contributors to noise pollution in the ocean (Military explosive testing is #1).

Vibroseis

A possible solution is marine vibroseis, which uses a diaphragm-like device to release the same amount of energy as an air gun but spread over a longer wavelength resulting in a deep hum, rather than a bang.

Force Blue, Inc.

ONE TEAM. ONE FIGHT.

Established as a 501c3 nonprofit organization in 2016, FORCE BLUE seeks to address two seemingly unrelated problems — the rapidly declining health of our planet's marine resources and the difficulty returning combat veterans have in adjusting to civilian life — through one, mission-focused program.

By uniting the community of Special Operations veterans with the world of marine science and conservation, FORCE BLUE has created a transpartisan model of caring, cooperation and positive change with the power to restore lives and restore the planet.

FORCE BLUE recruits the most highly skilled, highly trained individuals within the veteran Spec Ops community, retools them under the tutelage of esteemed marine scientists and conservationists, then redeploys them on missions of preservation and restoration around the globe. The result is a work force capable of achieving great things for themselves, their families, the communities FORCE BLUE serves and the planet as a whole.

"By starting a program that isn't just about helping veterans or just about helping the marine environment, but about helping both, we're really uniting two worlds." Says Ritterhoff. "Hopefully, FORCE BLUE will encourage each to look at the other with a bit more empathy and understanding."

Our capabilities include: Coral Restoration & Rehabilitation, Coral Disease Response, Marine Debris Removal, Underwater Survey Work, Shipwreck Mapping, Invasive Species Eradication, Marine Mammal & Reptile Rescue, Shark Tagging, and Scientific Research & Development.

"It feels good to actually take the training that we had and implement into real world use. That was one of the things I loved about the military. We had mission. We had purpose. There was something that we were doing every day and there was reason behind it." Geoff Reeves, Navy SEAL

We invite everyone to our website (www.ForceBlueTeam.org) for more information.

Quieter Propellers

Hakai Magazine explains that technologies such as low-noise propulsion systems—which use larger and slower-spinning propellers to minimize cavitation—and vibration-isolated mounting systems that get ship engines off the engine room floor, have allowed the National Oceanic and Atmospheric Administration (NOAA) to tone down the growl of some of their vessels.

These tactics can notch things down by 30 decibels, give or take, which translates to a 400 to 600% difference in the amount of sound animals hear. For researchers conducting stock assessments of fish, these increasingly stealthy ships are valuable.[289]

Reduced Speed

The U.S. National Park Service sites a study which determined that cruise ship speed was the dominant factor affecting how much noise whales were exposed to in Glacier Bay. In fact, the median *cumulative sound exposure level* (CSEL) values from two "slow" ships were lower than from just one "fast" ship. On days with fewer or slower cruise ships, the noise from the three daily tour vessels, which are much smaller than cruise ships, was more important to the bay's total underwater soundscape.

Another practical and important finding was that even though slower cruise ships produce longer exposure times (that is to say, they pass by more slowly), they produce much lower CSEL than faster cruise ships. The difference was substantial. The "whales" in simulations with cruise ships traveling at 13 knots were exposed to CSELs that were three times lower than when ships traveled at 20 knots. Even in cases where the ship was only a few decibels quieter at a slower speed, CSELs were lower,

---

[289] Hakai Magazine: Commercial Ships Could Be Quieter, but They Aren't
https://www.hakaimagazine.com/news/commercial-ships-could-be-quieter-they-arent/

even though the ship's transit past the "whale" takes much longer.[290]

The solutions mentioned above are those which will require your diligent stewardship and support in order to create regulations for the purpose of protecting our vulnerable marine species.

Thermal Pollution

As we discovered in our chapter on the ocean, it is the thermal expansion of water that is the cause of rising sea levels. This makes the topic of thermal pollution a crucial practice to eliminate. To remind you, thermal pollution is the practice of removing water from a natural source, circulating it through machinery to cool it down, and then ejecting that heated water back into the natural resource. Our research on solutions to the problem of thermal pollution lead us to a single article on *All About Water Filters* which offers 5 simple fixes:

1. Recycle used water.
   Whether on a small scale in your own home or on a larger scale at a local power plant, reusing water that has been heated for some other purpose already can go a long way toward cutting back on thermal pollution. For example, at a power plant, when water is heated from the cooling process, this water can be transferred to a generator or some other device and used to provider power and heat for other buildings. In your home, recycling your used water may be as simple as watering your plants with it, but it can still make a big difference in the long run.
2. Plant more trees.
   Planting trees anywhere is great for the environment, but when you work to plant trees along shorelines, you're helping cut back on erosion. Erosion can cause water pollution and can also lead to raising the temperature of the water due to storm runoff. Having more trees along the shoreline can help improve the ecosystem in

---

[290] National Park Service: Managing Underwater Noise Pollution
https://www.nps.gov/articles/canyouhearmenow.htm

the area overall, and can cut back on air pollution, too. There's no reason not to plant trees, so it can be a big help for your local water sources to organize a group to work on this activism task.

3. Spread awareness.
Hand out flyers, give talks and demonstrations, and even just tell your friends and family about your cause. There's no need to go over the top and be rude or unpleasant about it; just make sure everyone who knows you or comes into contact with you takes away something useful in terms of knowledge about thermal water pollution. And if you're trying to get some things changed locally where you live, be sure you reach out to companies, government officials, and other organizations that may be willing to help you. A little effort goes a long way!

4. Use cooling towers.
Power plants that use water as a coolant for their operations can look into using cooling towers that may help the thermal pollution situation. Simply put, a cooling tower allows the heated water to cool back down to the temperature of the water in nature before it is returned to its original source. This keeps it from raising the ambient water temperature at all, and it prevents the problem of thermal pollution in the environment altogether—at least where it's caused by power plant operations.

5. Use other cooling agents.
These power plants can also look into other sources of cooling, such as cooling agents and other methods as well. It can take a lot of activism to get an established power plant to change their cooling methods or agents, but when you work hard on something like this, chances are good your voice will eventually be heard. Don't be afraid to stand up and try to get some changes made when you can.

Please read the whole article here:

http://all-about-water-filters.com/solutions-to-thermal-pollution/

Ghost Nets

A wonderful organization called Oliver Ridley Project who are focused on the Indian Ocean, has the best explanation for *Ghost Nets* as well as solutions and opportunities for you to volunteer with them. This is directly from their website:

Ghost nets are commercial fishing nets that have been lost, abandoned, or discarded at sea. Every year they are responsible for trapping and killing millions of marine animals including sharks, rays, bony fish, turtles, dolphins, whales, crustaceans, and birds. Ghost nets cause further damage by entangling live coral, smothering reefs and introducing parasites and invasive species into reef environments.

In addition, ghost nets affect the sustainability of well-managed fisheries by damaging boats and killing species with economic value. They also impact the beauty of shorelines, resulting in expensive cleanup costs and financial loss for the tourism and diving industry. Olive Ridley Project volunteers have removed more than 1,400 ghost nets and recorded 812 trapped turtles: which make up 89% of the trapped turtles recorded (Numbers accurate as of July 2019).

The causes are multi-layered which means the solutions are as well. Please visit the Olive Ridley Project for a better understanding of causes, prevention and solutions.

https://oliveridleyproject.org/what-are-ghost-nets

Plastic

At present, it is nearly impossible to get through the day without touching something made of plastic. We already covered the downfalls of this petroleum-based material in *The Virus Human,* so now we can focus on solutions:

Plastic recycling

It alleviates so much guilt for us environmentally conscious humans when we toss that piece of plastic in the blue recycling bin. But what you need to know is that 90% of it can't be used. There are a number of reasons including the type of label or still present food matter. So, when it comes to recycling, the most effective method we have available is the process by which old plastic is chemically returned to its molecular beginnings. That means 95% of the plastic a chemical recycling plant receives can be processed and returned to its useable base. You can help reduce the amount of trash that ends up in our ocean by encouraging your local recycling management to use the innovative Agilyx method of chemical recycling. You can contact Agilyx for more information here:

https://www.agilyx.com/

Circular Economy

There are many uses for plastic that keeps it a necessary material, so until we can change the manufacturing of plastic over to more sustainable materials, (you'll see some below) we need to keep those things made of plastic in circulation and reuse them instead of making more. This can mean reusing end products that can be refurbished, or chemically recycling the components of plastic for reuse.

Eco-packaging

While it would be fantastic if we could all shop at farmer's markets and refill stations, it's not always possible with our complicated and fast-moving lives. While every day manufacturers are joining the ranks of CSR (corporate social responsibility) we consumers can choose the most sustainable of the packages to which we have access now. While our favorite solution is seed-bearing, plantable and compostable eco-packaging, is not available en masse yet; however, there are many paper and glass options already on the shelf. Talk to your grocers and request more sustainable options. Get that ball rolling and choose wisely grasshoppers. Here are a few awesome examples of eco-packaging:

Takeout containers:

https://www.goodstartpackaging.com/take-out-containers/

Eco-friendly shipping options:

https://www.ecoenclose.com/

Zero waste plantable packaging:

https://www.botanicalpaperworks.com/catalog/sustainable-eco-packaging

Edible packaging

Responsible manufacturers are providing sustainable packaging you can eat. This awesome article by *Foodtank* gives you a list of 18 of them:

https://foodtank.com/news/2018/09/have-your-food-and-eat-the-wrapper-too/

Single use alternatives

Eco-Products is a company providing alternative products; anything from cups to compostable bags:

https://www.ecoproducts.com

Bio-based alternatives

Hemp Plastic

Another great innovator is HempPlastic.com. This company provides hemp-plastic for manufacturing processes, from injection moldable options, to hemp films engineered for food to pharmaceutical and agricultural products. In addition, they also provide HempPropylene, HempEthylene, HempPLA, and HempABS. For the end user, they design and manufacture eco-friendly products customized to meet specific business needs from hemp plastic drink bottles to hemp plastic bags or straws and options for customized products.

https://hempplastic.com/

Coffee stuffs

AllThings.bio has a fascinating article about all the things our leftover coffee grounds can make. Thinks like:

- Coffee cups
- Fabric
- Furniture
- Lamps
- Ink
- Biofuels
- Water filtering
- Air filtering
- Filler for road surfaces
- Flower pots
- Pencils
- Cosmetics
- Jewelry
- Eyewear

Have that second cuppa Joe and read the whole article here:

http://www.allthings.bio/the-many-ways-of-turning-coffee-waste-into-valuable-materials-and-products/

We brilliant humans have come up with marvelous solutions to our worst problems, and although we have many alternatives to plastic within our reach, the important thing is that you begin to make changes in your daily living and shopping habits. For just a single day, every time you touch a piece of plastic, think of how you can make a sustainable choice based on what you are touching. Could this be fixed and reused, recycled or replaced with a more sustainable choice the next time? Is there another choice on the shelf that is more sustainable than this one? Use your power and save the world.

Sewage and Septic

When it comes to *our* stance on big solutions for a planet already experiencing human population stress, we believe every home and building should be a completely self-sufficient island. That means, manage your own waste. When the problems are ours to contend with, it's surprising what we are willing to change for convenience. Up to this point, we all *could* have been living zero-waste lives; we just didn't have a good enough reason to do it. Our reason is here. Our planet needs a break from our present way of life. As for the solution to sewer and septic? Our favorite solution is singular and simple. We are huge fans of incinerating toilets; specifically the Cinderella Company line.

An incineration toilet is a solution where biological waste is incinerated to ashes, completely free of bacteria. Cinderella Incineration Toilets do not pollute the environment and need no permissions for installation in sensitive areas.[291]

These toilets are beautiful, as toilets go. They are sleek, modern and completely self-sufficient. Yes, they are expensive; around $5,000, but when you consider you have no plumbing, sewage fees or septic costs, they will still bring your project out ahead financially. They are awesome travel solutions as well as daily living.

You need to know that most new build construction projects above a particular square footage in size, have plumbing requirements. Each municipality is different, so you 'll want to check your local laws before you rip out your toidy, but our solution to that problem is to live tiny and minimal.

Please visit Cinderella Company here and check out this, one of our

---

[291] Cinderella: Eco-friendly incineration toilets https://www.cinderellaeco.com/no-en/

favorite solutions:

https://www.cinderellaeco.com/no-en/

Species Extinction

The solutions for Species Extinction mostly fall into the category we referenced earlier as **leave half of it alone!** Unless you can help remove the disturbance (generally pollution) and rebuild an ecosystem, don't disturb it further. Don't poke it. Don't goose, jab, prick, prod, punch, push or shove it. And especially, don't kill its predators. Where possible, let it repair itself on its own.

There's a reason the phrase "let nature take its course" exists: New research done at the Yale University School of Forestry & Environmental Science reinforces the idea that ecosystems are quite resilient and can rebound from pollution and environmental degradation. Published in the journal PLoS ONE, the study shows that most damaged ecosystems worldwide can recover within a single lifetime, if the source of pollution is removed and restoration work done.

The analysis found that on average forest ecosystems can recover in 42 years, while in takes only about 10 years for the ocean bottom to recover. If an area has seen multiple, interactive disturbances, it can take on average 56 years for recovery. In general, most ecosystems take longer to recover from human-induced disturbances than from natural events, such as hurricanes.

To reach these recovery averages, the researchers looked at data from peer-reviewed studies over the past 100 years on the rate of ecosystem recovery once the source of pollution was removed.

Interestingly, the researchers found that it appears that the rate at which an ecosystem recovers may be independent of its degraded condition: Aquatic systems may recover more quickly than say, a forest, because the species and organisms that live in that ecosystem turn over more rapidly than in the forest.[292]

Consequently, one of the most important actions an average citizen can take is to support the Endangered Species Act and government representatives who also support a healthy biodiversity in our environment.

The law was inspired, in large part, by the bald eagle. In 1966, concern for our national bird—which had drastically decreased in population due to hunting, habitat loss, and the rampant use of the toxic pesticide DDT—motivated Congress to pass the Endangered Species Preservation Act. It stated that the U.S. Departments of Interior, Agriculture, and Defense must protect listed species and their habitats. In 1973, after a series of amendments, this original framework expanded and evolved into the Endangered Species Act.

The act doesn't just stop the bleeding; it requires the federal government to prepare a recovery plan so that the listed species can be restored to a healthy population—and eventually come off the list. A whopping 99% of the species granted protection under the act have managed to survive until today, and a growing number—including the bald eagle, American peregrine falcon, Eggert's sunflower, and red kangaroo—have recovered enough to be delisted, meaning they're no longer in danger.

Saving one species can save countless others. Because each plant or animal is part of a larger ecosystem, preserving any one could create a ripple effect. The gray wolf is a great case study. When the species was reintroduced to Yellowstone National Park—more than 20 years after being listed as endangered in 1974—the impact was far-reaching. The packs helped keep the elk population in check, which meant that willow and aspen trees were in less danger of being overeaten. The branches and leaves of those trees cooled the streams, which boosted the population of

---

[292] TreeHugger: Good News: Most Ecosystems Can Recover in One Lifetime from Human-Induced or Natural Disturbance https://www.treehugger.com/natural-sciences/good-news-most-ecosystems-can-recover-in-one-lifetime-from-human-induced-or-natural-disturbance.html

native trout, provided homes for migratory birds, and supplied more food for beavers. The dams built by the beavers created happier marshland habitat for otters, mink, and ducks. And the benefits go on and on.

Climate change is making the act even more important. In 2008, the polar bear became the first species given protection under the Endangered Species Act due to the threat of global warming (melting sea ice, in this instance). In 2011, the whitebark pine became the first widely dispersed tree species to be designated as a candidate for endangered species protection, as it was threatened by pine beetles spreading to higher elevations due to warmer temperatures.

Congress consistently tries to weaken its own law. Bills are regularly introduced to undermine the act. See the "Endangered Species Management Self-Determination Act," "Common Sense in Species Protection Act of 2015," and the "21st Century Endangered Species Transparency Act." Much of this proposed legislation places short-term economic gain above long-term conservation efforts, and demands changes (requiring state consent, for example) that would make it much more difficult to protect species. And sometimes other less-obvious bills—relating to the federal budget, defense, or food or water security—can have provisions that chip away at protections for specific species.

Businesses also try to weaken or eliminate the act. Take the case of the coastal California gnatcatcher, a small gray bird in Southern California. Real-estate developers and toll road agencies have fought federal protection for this particular species, which picked a piece of pretty pricey land on which to build its nests. Some corporations have even filed a petition claiming the bird isn't a valid subspecies to try to delist it.[293]

It is extremely important to elect lawmakers who support the Endangered Species Act. Ecosystem advocates will help support/strengthen its reach and coverage. Writing emails/letters, making telephone calls and

---

[293] NRDC - 10 Things You Didn't Know About the Endangered Species Act
https://www.nrdc.org/stories/10-things-you-didnt-know-about-endangered-species-act

conducting in-person conversations with current congressmen and women are all effective ways to have your voice heard.

Trophy hunting is the shooting of carefully selected animals – frequently big game such as rhinos, elephants, lions, pumas and bears – under official government license, for pleasure. The trophy is the animal (or its head, skin or any other body part) that the hunter keeps as a souvenir.

Some countries allow a small number of endangered species to be killed in the wild by sports hunters and, with approval from the Convention on International Trade in Endangered Species (CITES), it is still possible to take the trophies home. Allowing endangered species to be killed for sport is counterintuitive. Poachers are slaughtering about 100 elephants and 3–4 rhinos every day, for example, so allowing trophy hunters to kill yet more seems absurd. Critics are also concerned about the mixed messages it sends local people: they can't hunt endangered species, but rich Westerners can.

The USA legally imports no fewer than 126,000 animal trophies every year, and the EU some 11,000–12,000 (representing 140 species, including everything from African elephants to American black bears), not counting those trophies taken in the countries themselves.[294]

Trophy hunters fork over huge payments for the pleasure of shooting an animal. Critics say most of this money goes to the hunting elite and corrupt government officials. Perhaps 3% of the money goes toward any semblance of maintaining the species. Here is this real problem. An animal can only be killed once and it is gone forever, whereas a single tiger or rhino could potentially earn money from traditional ecotourism for many years in the future.

As we have seen in the study above, endangered species can correct their own ecosystem within a human lifetime when the ecological disturbance has been removed. Put down the trophy hunting rifles and poaching

---

[294] Discover Wildlife: An introduction to trophy hunting
https://www.discoverwildlife.com/animal-facts/an-introduction-to-trophy-hunting/

toolkits, and we can save these species for future generations.

Speaking of poaching, the taking of life for profit is closely related to trophy hunting. The trophy hunter takes souvenirs, head and fur for ego enrichment, while the poacher takes only the valuable parts of the animal for financial enrichment. In both cases, the animal is dead.

Poaching of any animal is illegal. Period. It usually occurs when an animal possesses something that is considered valuable (i.e. the animal's fur or ivory).[295] Here are a few facts about the perceived value of animal parts:

Many countries believe that the rhino horn is an important ingredient for many medicines. This is false. Rhino horn has the same medicinal effect as chewing on your fingernails - aka none.[296]

Tigers are primarily killed to supply underground black markets with its organs, pelts, and bones. These items are highly regarded in eastern medicine (although these treatments have been disproved and have no real medical value).[297]

Typically the largest adults, with the biggest tusks are poached – putting the matriarchs of elephant herds at the greatest risk.[298] This of course does the most damage to the elephant ecosystem when they are left without their leader.

Legal hunters kill tens of millions of animals per year. For each of those animals, another animal is illegally killed.[299]

---

[295] Simonetta, Alberto. "CONTROL OF POACHING AND THE MARKET FOR PRODUCTS SUCH AS IVORY, RHINO HORN, TIGER AND BEAR BODY PRODUCTS." Department of Animal Biology and Genetics "Leo Pardi", University of Florence, Florence, Italy. Web Accessed March 28, 2015.
[296] African Wildlife Federation. "Africa's Poaching Crisis." Web Accessed March 28, 2015.
[297] World Wildlife Fund. "Save Tigers Now." Web Accessed April 4, 2015.
[298] University of Washington. "Effects of poaching on African elephants." Web Accessed March 28, 2015.
[299] Fact Retriever: 60 Tragic Poaching Facts

Conservationists estimate that between 30,000 and 38,000 elephants are poached annually for their ivory.[300] Since there may be as few as 400,000 elephants left, this problem will be at a very severe DEFCON 1 emergency action within the next few years.

Again, what can one person do now to help?

Support anti-poaching organizations that are on the front lines fighting the good fight, like Veterans for Wildlife. Their 5-year strategic plan is centered on empowering veterans from around the globe and preventing wildlife crime, predominantly across Southern Africa which is home to approximately 93% of the world's white rhino and 40% of its black rhino. Find out more about this fantastic group at:

https://www.veterans4wildlife.org

The International Anti-Poaching Foundation (IAPF) builds and leads large scale conservation operations, maintaining a strong focus on community development in order to conserve biodiversity. Anti-poaching rangers form the first and last line of defense for nature.

https://www.iapf.org

World Wildlife Fund (WWF) recently used a grant from Google.org to engineer a remarkable new thermal and infrared camera and software system that can identify poachers from afar and alert park rangers of their presence.

https://www.worldwildlife.org

Waste Management

The single greatest hope we have of managing our massive barges of waste is to stop generating it and becoming a zero waste species. We're a

---

https://www.factretriever.com/poaching-facts
[300] Fact Retriever: 60 Tragic Poaching Facts
https://www.factretriever.com/poaching-facts

ways off but the quickest way to get there is to support a *Circular Economy*. The best way to answer "What is a Circular economy" is to ask, "Does a bear poop in the woods?" Of course she does. The poop fertilizes plants, plants produce food and oxygen for the bear, and that my friend, is a circular economy. A magnificent organization out there doing great things for a circular economy is the Ellen MacArthur Foundation:

A circular economy is based on the principles of designing out waste and pollution, keeping products and materials in use, and regenerating natural systems.

Up until recently, we humans have lived in a linear economy: Take, Use, Dispose, and Repeat. This way of life has created most of the problems we have addressed in this book. Each time we toss out the old thing, we have wasted the resources used to create it and created a burden of waste and toxic waste by-products for another ecosystem. By creating a more natural relationship that works with our planet's precious resources, the model becomes a circular way of life.

Design

By redesigning the products and components we use, we can create them in such a way that we build our natural resources instead of depleting them. Even the packaging can create a positive economy. A company called *Botanical Paperworks* sells packaging that can be planted instead of throwing it away. You can literally plant the package and grow flowers, herbs, or veggies (BotanicalPaperworks.com). And *Saltwater Brewery* created 100% biodegradable six pack rings that can be eaten by sea creatures (Saltwaterbrewery.com).

Recirculate

We have generated mountains of used appliances, electronics, clothing, shoes and vehicles. Many companies are now retaining ownership of these items and leasing them to us for a fee. That way, the broken or worn-out product returns directly to the manufacturer, who in turn, reuses those base materials. Yes, even clothing. There's a company we mention in *Plastic* solutions called *Rent the Runway* that offers this very service. You can pay between $30 and $159 to have access to a constantly rotating wardrobe. They even have a program for kids and curate for you based on your style (RentTheRunway.com).

Regenerate

By using methods and systems that mimic nature, we renew the depletion of natural resources. We can make products that plant seeds, nourish soil and water, feed other Earthlings, and even return our own biological waste and our bodies to the Earth in such a way that it feeds the Earth. This is the full circle of a circular economy.

The *Triple Pundit* profiled a few businesses already practicing these sustainable habits[301]:

Levi Strauss

Levi Strauss & Co. is no stranger to sustainability. From utilizing sustainable raw materials to reducing water waste, the company keeps its green cred sky-high. Now, it's inching ever closer to a closed-loop supply chain.

Levi Strauss and its retail chain Levi's also help customers reduce their own footprints by accepting clothing and shoes of all brands for recycling. Through its recycling partner, I:CO, clothing is sorted for resale, reuse and recycling — ensuring nothing goes to waste.

Nike

Nike continues to increase the sustainability of its supply chain. That includes huge strides in waste reduction. A whopping 71% of Nike

---

[301] Triple Pundit: 3p Weekend: 7 Companies Making the Circular Economy a Reality
https://www.triplepundit.com/story/2016/3p-weekend-7-companies-making-circular-economy-reality/22376

footwear is made with materials recycled from its own manufacturing process. And the brand recovered 92% of its trash last year.

"By creating low-impact and regenerative materials, we can continue to move toward a high-performance, closed-loop model that uses reclaimed materials from the start," Mark Parker, Nike's president and CEO, wrote in its 2015 CSR report. "Coupled with smarter designs, we can create products that maximize performance, lighten our environmental impact and can be disassembled and easily reused".

Dell

It's hard to cover the circular economy without mentioning Dell. The electronics company is outspoken about its quest for a closed-loop supply chain, and each year it only grows closer to this goal.

Last year, the company took another significant step forward with its OptiPlex 3030 All-in-One computer. The model contains at least 10% repurposed plastic from recycled electronics, and set a new closed-loop standard for the industry.

A few months later, the company announced it would do even more to boost plastics recycling and the reclamation of carbon fiber materials. The revamp includes 35 products, and will recapture millions of pounds of plastic and carbon fiber material, Dell said. The company already recycles plastic components from over 30 flat panel-monitor models and three desktop models.

Water

The best solution we have to the global water crisis is to stop wasting the stuff. We must end the unsustainable practices wasting the water we have, and that means you must take a stand. Regulations protecting all Earthlings and precious finite resources need to be enforced and strengthened, not relaxed. Use your voice. Use your power and make corporations and governments accountable for the flagrant misuse of resources.

On a smaller scale, there are many things you can do in your daily life to protect our precious water:

- Use an Energy Star/energy efficient dishwasher instead of washing by hand. This can save almost 24 gallons of water per load
    - You can also clean your dishes with wood-ash paste (ash with a bit of water) just wipe clean after a scrub
- Turn off the faucet while brushing your teeth
- Only run the washing machine and dishwasher when you have a full load
- Use a low flow shower head and faucet aerators
- Fix leaks
- Install a dual flush or low flow toilet, or put a conversion kit on your existing toilet
- Don't overwater your lawn or water during peak periods, and install rain sensors on irrigation systems
- Install a rain barrel for outdoor watering
- Plant a rain garden for catching storm water runoff from your roof, driveway, and other hard surfaces
- Monitor your water usage on your water bill and ask your local government about a home water audit
- Share your knowledge about saving water through conservation and efficiency with your neighbors[302]

On a larger scale, remember that everything you use washes down the watershed. There are some very old and natural alternatives for cleaning our bodies and our stuff…use those; but there are also some new technologies available to help:

DIY Cleaning Solutions

WellnessMama.com gives us a great list of natural cleaning items to keep

---

[302] American Rivers: 10 WAYS TO SAVE WATER AT HOME
https://www.americanrivers.org/rivers/discover-your-river/top-10-ways-for-you-to-save-water-at-home/

on hand. Go to the site for the more information and link to glass spray bottle:

- White vinegar
- Liquid castile soap or Sal Suds
- Natural salt
- Baking soda
- Borax
- Washing soda
- Hydrogen peroxide
- Lemons
- Microfiber cloths
- Essential oils (optional)
- Aa spray bottle or two (preferably glass)

For more information, visit:

https://wellnessmama.com/6244/natural-cleaning/

Natural Body Care

As we have curated information for you from experts in their field, we found a wonderful resource to tell you all the natural things you can use for body care. Betsy Jabs of DIYnatural.com has a great recipe for body wash you can find here:

https://www.diynatural.com/homemade-body-wash/

Natural Shampoo

WellnessMama.com has another great recipe here to make your own natural shampoo:

https://wellnessmama.com/3701/homemade-shampoo/

One-Stop Shopping

If you're not the DIY type, you can find some wonderfully natural home products here:

https://www.naturalhomebrands.com/pages/about-our-products

Electrolyzed Water

Electrolyzed water is produced by the electrolysis of ordinary tap water containing dissolved sodium chloride which produces a solution of hypochlorous acid and sodium hydroxide. The resulting water is a known cleanser and disinfectant / sanitizer. Electrolyzed water is an all-natural, non-toxic, and non-hazardous solution that can be used for disinfecting or degreasing. As a disinfectant, it is a powerful oxidant that is 100 times more powerful than chlorine bleach.

There are units big enough for medical and industrial use, but for you we recommend the home version by Ecolox Tech:

https://www.ecoloxtech.com

When it comes to water, the whole goal is to protect what we have, keep it clean and within the reach of every single Earthling. You met these guys in *Energy* but we are mentioning *Off Grid Box* again here because they have created a way to make water within the reach of the most remote and desert villages. They are presently working with NGOs, engineering firms, and development agencies, but they are proving that the technologies to save every citizen are within our reach.

Off Grid Box

- **Water collection system**: untreated water can be collected by the integrated rain capture system or from external sources such as a well, river, ocean or external tank.
- **Water storage tank**: the large integrated polyurethane tanks stores the untreated water to be cleaned when needed.

- **Water treatment system**: some of the electricity is used to filter and sterilize the untreated water and then distribute the clean water when needed

https://www.offgridbox.com/

Water from The Sun

Okay, so it's not directly from the sun, but a very cool company from Arizona called *ZEROMASS Water* created a way to extract water from the air using thermodynamics. Two of their solar "hydropanels" can mine enough water from air to fill the equivalent of as many as 600 bottles of water a month. That's enough for 6 people. This off-grid water source can even extract water in humidity as low as 10%. We consider this innovation a must for anyone heading into our questionable future with relation to water, ownership and clean sources.

https://www.zeromasswater.com/na/

Air

Until we get a firm grasp on the cojones of the industrial machine, we are at their mercy. This is a global issue as the air in the U.S. and India will blow to all four corners of the world. On a small scale you can keep air purifiers around and in the event of a catastrophe, fire or chemical spill; we recommend that every family has gas masks in your survival kits for every member of the family, including pets. Be prepared for anything, as we learned from the Paradise fire.

The big-fix solution to poor air quality is *regulation*. Be aware of your local statutes and restrictions on industrial air pollution and make sure your family is protected by continuing to improve them for the long-term. Find a local watchdog organization to join or create one to make certain your air is up to snuff, and if it isn't, organize and demand improvements.

Every nation has its own set of challenges, and following this month's

UN Global Climate Assessment, many countries are taking rapid action to correct our mistakes with regard to the health of our planet. In the U.S., we have the beginnings of an effective regulatory construct called the *Clean Air Act*, but recent actions by the Trump administration have set in motion roll-backs to 85 of these protections:

		53 ROLLBACKS COMPLETED	32 ROLLBACKS IN PROCESS	85 TOTAL ROLLBACKS
	Air pollution and emissions	10	14	24
	Drilling and extraction	9	9	18
	Infrastructure and planning	12	1	13
	Animals	9	1	10
	Toxic substances and safety	4	1	5
	Water pollution	5	2	7
	Other	4	4	8

Image: https://www.nytimes.com/interactive/2019/climate/trump-environment-rollbacks.html

That's the wrong direction for where we are headed. Make your voice heard on every level of your government's infrastructure and protect the future for our planet and all those who rely on her good health for theirs.

Indigenous Living

There are more than 300 million indigenous people in virtually every

region of the world, including the Sámi peoples of Scandinavia, the Maya of Guatemala, numerous tribal groups in the Amazonian rainforest, the Dalits in the mountains of Southern India, the San and Kwei of Southern Africa, Aboriginal people in Australia, and, of course the hundreds of Indigenous Peoples in Mexico, Central and South America, as well as here in what is now known as North America.

There is enormous diversity among communities of Indigenous Peoples, each of which has its own distinct culture, language, history, and unique way of life. Despite these differences, Indigenous Peoples across the globe share some common values derived in part from an understanding that their lives are part of, and, inseparable from the natural world.

Onondaga Faith Keeper Oren Lyons once said, "Our knowledge is profound and comes from living in one place for untold generations. It comes from watching the sun rise in the east and set in the west from the same place over great sections of time. We are as familiar with the lands, rivers, and great seas that surround us as we are with the faces of our mothers. Indeed, we call the Earth Etenoha, our mother from whence all life springs".

Indigenous people are not the only people who understand the interconnectedness of all living things. There are many thousands of people from different ethnic groups who care deeply about the environment and fight every day to protect the Earth. The difference is that indigenous people have the benefit of being regularly reminded of their responsibilities to the land by stories and ceremonies. They remain close to the land, not only in the way they live, but in their hearts and in the way they view the world. Protecting the environment is not an intellectual exercise; it is a sacred duty. [303]

Indigenous groups view their role on Earth as a spiritual responsibility to preserve all life for future generations; plants, animals, water, air, even

---

[303] Cultural Survival: BEING INDIGENOUS IN THE 21ST CENTURY
https://www.culturalsurvival.org/publications/cultural-survival-quarterly/being-indigenous-21st-century

the Earth itself as a living being. There is also a strong sense of cooperation among indigenous peoples that all tribal members must participate in these preservation activities. They are taught as part of indigenous culture and spirituality that should be welcome in any modern lifestyle.

In our research to understand more about cooperative village lifestyles, we took a very close look at the beliefs and spirituality of our indigenous brothers and sisters. Because the notion of modern societies has only been around for about 200 years, we wanted to know more about how people lived before that time, and since, to learn how average humans can begin to more efficiently use food, water, waste and energy. It turns out that indigenous living is the best example of sustainability remaining today.

"Living well is all about keeping good relations with Mother Earth and not living by domination or extraction," says Victoria Tauli Corpuz, the UN Special Rapporteur on the Rights of Indigenous Peoples.

We found that smaller village living provides the best opportunity to reduce the human carbon footprint - the amount of CO_2 and other carbon compounds emitted due to the consumption of fossil fuels by a particular human or group. Everyone in the community contributes to the community and the community provides for all in a common goal of sustainability. It's called teamwork.

Tribal governments in the United States today, for example, *exercise a range of sovereign rights. Many tribal governments have their own judicial systems, operate their own police force, run their own schools, administer their own clinics and hospitals, and operate a wide range of business enterprises. There are now more than two dozen tribally controlled community colleges. All these advancements benefit everyone in the community, not just tribal people. The history, contemporary lives, and future of tribal governments are intertwined with that of their neighbors.*[304]

We can also learn a great deal from indigenous farming practices:

- Agroforestry involves the deliberate maintenance and planting of trees to develop a microclimate that protects crops against extremes. Blending agricultural with forestry techniques, this farming system helps to control temperature, sunlight exposure, and susceptibility to wind, hail, and rain. This system provides a diversified range of products such as food, fodder, firewood, timber, and medicine while improving soil quality, reducing erosion, and storing carbon.
- The principles of crop rotation have been successfully used for thousands of years in agriculture and are still used today. Crop rotation is the practice of growing different crops on the same land so that no bed or plot sees the same crop in successive seasons. It is a practice designed to preserve the productive capacity of the soil, minimize pests and diseases, reduce chemical use, and manage nutrient requirements, all of which help to maximize yield. The practice of crop rotation builds better soil structure and increases the ability to store carbon on farms.
- Mixed cropping, also known as intercropping, is a system of cropping in which farmers sow more than two crops at the same time. By planting multiple crops, farmers can maximize land use while reducing the risks associated with single crop failure. Intercropping creates biodiversity, which attracts a variety of beneficial and predatory insects to minimize pests and can also increase soil organic matter, fumigate the soil, and suppress weed growth.
- Polyculture systems involve growing many plants of different species in the same area, often in a way that imitates nature. By increasing plant biodiversity, polyculture systems promote diet

---

[304] Cultural Survival: BEING INDIGENOUS IN THE 21ST CENTURY
https://www.culturalsurvival.org/publications/cultural-survival-quarterly/being-indigenous-21st-century

diversity in local communities, are more adaptable to climate variability and extreme weather events, and are more resilient to pests and diseases. Polycultures are integral to permaculture systems and design and provide many advantages such as better soil quality, less soil erosion, and more stable yields when compared to monoculture systems.

- Water harvesting is defined as the redirection and productive use of rainfall, involving a variety of methods to collect as much water as possible out of each rainfall. Many water harvesting structures and systems are specific to the ecoregions and culture in which it has been developed. This may involve collecting water from rooftops, from swollen streams and rivers during monsoon season, or from artificially constructed catchments. This ensures that farmers have a substantial amount of water stored up in the case of drought or limited rainfall.[305]

Only about 1,200 new drugs have been approved by the U.S. Food and Drug Administration (FDA) since 1950.[306] Because of the many side effects attributed to some of these chemical compound remedies, microbial resistance and honestly a sometime lack of cure, many traditional indigenous curative formulas are making a comeback as "conventional" medicines begin to loosen their hold on the pharmaceutical industry.

For thousands of years, traditional indigenous medicine has been used to promote health and wellbeing for millions of Native people who once inhabited this continent. Native diets, ceremonies that greet the seasons and the harvests, and the use of native plants for healing purposes have been used to promote health by living in harmony with the Earth. Today, Native Americans frequently combine traditional healing practices with allopathic medicine to promote health and wellbeing. Ceremony, native

---

[305] Resilience: Five Indigenous Farming Practices Enhancing Food Security https://www.resilience.org/stories/2017-08-14/five-indigenous-farming-practices-enhancing-food-security/

[306] Munos B. Lessons from 60 years of pharmaceutical innovation. Nature Reviews Drug Discovery. 2009;8(12):959–968.

herbal remedies, and allopathic medications are used side by side. Spiritual treatments are thus an integral part of health promotion and healing in Native American culture.[307]

These and other practices work equally well in smaller villages, neighborhoods and intentional communities to help make life better and lighter on the Earth at the same time. Ecovillages, as you will see, are locally owned, socially conscious communities using participatory ways to enhance the spiritual, social, ecological and economic aspects of life.[308]

"This is all about finding ways for humanity to survive. Much of this is a return to the values and practices of indigenous peoples," says Lee Davies, a board member of the Global Ecovillage Network (GEN).[309]

We will include most of the solutions in this chapter as we discuss sustainable village living. I will leave you with one final thought from my Native American ancestors:

We will be known forever by the tracks we leave. – Dakota Proverb

Our Village

Our intent, up to this point, has been to take you with us on our journey to sustainable freedom and help you to envision your own utopia. Our village is our vision. Hopefully, we have provided the tools for you to envision yours as well. While we envision a life apart from the buzz of machines and infrastructure, we have family members and loved ones who thrive on the fire of city living and we have a great template for urban dwellers as well. Today, 55% of humans live in urban areas and

---

[307] PMC: Indigenous Native American Healing Traditions
https://www.ncbi.nlm.nih.gov/pmc/articles/PMC2913884/
[308] Our World: Living the Indigenous Way, from the Jungles to the Mountains
https://ourworld.unu.edu/en/living-the-indigenous-way-from-the-jungles-to-the-mountains
[309] Global Ecovillage Network https://ecovillage.org/

that number is expected to rise to 68% by 2050.[310] Yet, presently, there are massive food deserts in these urban regions; meaning, the citizens do not have local access to fresh produce. Our favorite solution for urban eco-villages is a vertical urban plan designed by *Urban Eco Villages* which includes massive vertical, hydroponic gardens amidst office and living space and solves all the big problems of energy and waste. These experts in their field aptly demonstrate their designs and solutions at:

https://ecovillage.org/urban-ecovillages-north-america

We, however, prefer a more Thoreauvian approach and envision our eco-village within a more natural setting. Below is our concept of an off-grid village. To reverently paraphrase the Marine Corp Rifleman's Creed[311] of my father, "This is our village. There are many like it, but this one is ours…". We are presenting one iteration of village living. The village you creatively design using these solutions or others should meet your group needs.

Begin your design after you have found suitable land to support your group needs. It should preferably include arable land for farming and a clean water source, though neither is required with our solutions. Be sure to check state and local requirements for clearing of land, permits, and constructing off-grid communities.

Every family, group, or individual will require personal living space. The tiny home movement has grown tremendously over the last few years. The demand for living small is getting bigger. More than half of Americans would consider living in a home that's less than 600 square feet, according to a survey done by the National Association of Home

---

[310] United Nations: Department of Economic and Social Affairs
https://www.un.org/development/desa/en/news/population/2018-revision-of-world-urbanization-prospects.html
[311] MarineParents: My Rifle: The Creed of a U.S. Marine
https://www.marineparents.com/marinecorps/mc-rifle.asp

Builders. And among Millennials, interest increases to 63%.[312]

After physical orientation plans and suitable designs have been created, there are several ways to approach the construction of your tiny homes. If you are a builder or someone with building skills, you may consider constructing your own small domicile. Klein (http://www.liveklein.com/) will soon offer a tiny home kit the buyer can build for himself or herself. An average 150-square-foot tiny home is a low-impact, off-the-grid structure made of wood, glass and canvas that can be carried into remote areas and constructed by hand.

You can find some of this year's best tiny home kits here:

The 9 Best Tiny House Kits of 2019

https://www.thespruce.com/best-tiny-house-kits-4691381

If you are like most of us, you may be overwhelmed with this whole process and need some help with it. There are hundreds of tiny home builders across the United States and more are getting into the business every year. Check for builders in your area.

Here are a few of our favorites, in no particular order and without endorsement:

Tiny Heirloom:

https://www.tinyheirloom.com/

Renu Foundation:

https://www.renufoundation.org

Search here for a tiny home builder near you:

https://www.homebuilderdigest.com/the-best-tiny-home-builders-in-all-50-states/

---

[312] CNN Business: Demand for tiny homes is getting bigger
https://www.cnn.com/2018/11/02/success/tiny-homes/index.html

Our village construction will take advantage of structural insulated panels (SIPs) as a high-performance building system for residential and light commercial construction. The panels consist of an insulating foam core sandwiched between two structural facings, typically oriented strand board (OSB). SIPs are manufactured under factory-controlled conditions, and can be fabricated to fit nearly any building design. The result is a building system that is extremely strong, energy efficient and cost effective. Building with SIPs will save you time, money and labor.[313]

SIP panels are a bit more expensive initially but if factoring in the labor savings resulting from shorter construction time and less jobsite waste, the end costs are about the same. Other savings are realized because smaller heating and cooling systems are required with SIP construction.

An essential aspect of any community that fosters cooperation and social networking is to have a common community building - the physical spaces and services that help meet the needs of the community as a whole. Common buildings can have a huge impact on the development of social networks and the subsequent values of community participation and civic engagement. For example, in our village, common spaces where members of the community can interact (containing personal remote work spaces, cooking facilities, child care space, meeting/party areas, hydroponics/aquaponics, Wi-Fi, etc.) will help us build and sustain social networks in your community.

It is important to note here that our community center will house an extensive hydroponic/aquaponics facility to grow fresh foodstuff for our community. Each member of the group will donate time to maintain the gardens as well as other areas of the community center.

If your community is looking for an all-in-one solution for power and water, you may want to consider the OffGridBox, mentioned earlier, which is a cost-effective way to establish communications and use solar energy to purify water and distribute energy.

---

[313] SIPA: What Are SIPs? https://www.sips.org/about/what-are-sips

The OffGridBox is an all-in-one container equipped with solar panels and a water purification system to enable homes and communities in remote areas to become sustainable and resilient. The company has already deployed a total of 38 off grid "boxes" globally. Around the world, millions of people lack electricity and clean water, and OffGridBox aims to bridge this gap by providing these essential services at an affordable price. With invaluable contributions from impact investors, they are able to facilitate economic growth, stimulate social development, empower women, and improve local health conditions.

What makes OffGridBox truly unique is the development of an advanced remote monitoring software to optimize water and power control. This sustainable competitive advantage includes sensors, algorithms, automated safety precautions, performance insights, automated energy recovery, and prioritization for appliances. With this data, not only can the box predict and automatize maintenance, but it can also serve as the most efficient system in the rural off-grid market. Overall, the remote monitoring technology simplifies the process for installers, operators, and end users.

One 6-foot cube provides purified water and electricity for 3,000 people, powered by the sun, run by women, and good for everyone. With an innovative business model, we provide affordable water and power for those living on less than $2 a day.

OffGridBox is the brainchild of founders Emiliano Cecchini and Davide Bonsignore from 2014. They became inspired to create more efficient technical solutions after working on several OXFAM clean energy projects in South Africa that took more than 3 weeks to install. Since then, they got it down to a 3 hour installation with more than 38 boxes deployed in 10 countries, serving thousands of people in difficult circumstances. Projects range from disaster relief, cooperatively with NGOs, to rural electrification. OffGridBox is headquartered in Boston, Massachusetts, at Greentown Labs and the company's mission is to provide affordable clean water and electricity in remote areas. For more information about OffGridBox, please visit www.offgridbox.com.

Our village will utilize a more distributed model for solar, water and communications for additional flexibility. We will be using hydropanels from *ZEROMASS Water* (https://www.zeromasswater.com/) for water extraction from the air at a rate of approximately 10 liters per day per panel. The SmartFlower solar technology will be used for its ability to track the sun's movements to maximize daily power collection. And finally, Hughes.Net and Viasat satellite communications are our favorite providers as rates fluctuate according to data needs.

We hope our vision has inspired your own visions of life in harmony with the natural world and provided viable choices to make your long-term dreams of happiness just a little bit closer to a genuinely attainable reality.

There are so many things I want to say to you but like I told you in the first chapter, this is the beginning of our conversation. I want to hold you and promise you it will be okay. My loves, please feel my Amazonian arms wrapped around you. It's going to be okay. We just had a raucous teenage hedonism binge, its 11:55 pm and the adults will be home at midnight. Let's get this beautiful home of ours cleaned up in this, the eleventh hour. We can do it. I promise. We got this.

When my oldest daughter was almost 2, we took a city bus from Reston, Virginia into the District of Columbia. On the way, she was gazing up at a beautiful blue sky and with all seriousness, turned to me and said, Mommy, do you think Heavenly Father is as handsome as Mr. Rogers? The frequency with which my childhood was disrupted and fatherless, Mr. Rogers was my constant. Having the tremendous good fortune to have my child's life equally as impacted by such a great man, I wish to share our love for him with you. Therefore, these are the words with which I will end:

We are all called to be 'Tikkun Olam,'

repairers of creation.

Mr. Fred Rogers

143

Epilogue

The Earth that we live in is miraculous. Humankind started showing up on Earth in the Middle Paleolithic period well over 200,000 years ago, spreading from Ethiopia to Parts Unknown, and the Earth didn't even blink from the actions taken by these early humans. Like-minded humans evolved into clans, which evolved into kingdoms, and later into nations, and the Earth continued her rotation as if nothing had changed for many millennia.

Later (sometime in the 1760s), the Industrial Revolution began, where there was a transition to newer manufacturing processes that used machinery and chemicals. With this revolution, there was also an unprecedented spike in the rate of population growth, and Mother Nature took notice. You see, the initial Industrial Revolution was powered by an immense amount of coal being burned (replacing other energy sources like wood, water and wind). This change caused an unparalleled amount of air pollution. Ever hear of London Fog? This was no mere clothing manufacturer or a tea drink, but referred to soup-like fog in London that originated from the excessive coal combustion.

This revolution was a major turning point for Earth's ecology and humankind's relationship and effects with Earth's environment. The environmental effects continued to intensify with the addition of fossil fuels and combustion (oil drilling, offshore oil drilling, fracking, etc.), mining, and an exponential population growth pattern without much thought or notice, until the 1960's. By then, degradation of the air, water, and soil via nearly unfettered polluting was so bad that the Cuyahoga River in Ohio actually caught ON FIRE at least a dozen times!

Ultimately it took just one person to wake the human population up to realize that we have been slowly killing the planet for nearly 200 years (and the species that live on it both human and non-human alike) – Rachel Carson and her book *Silent Spring*. In that book, she eloquently displayed the cause and effect of humankind's actions on the environment due to the Industrial Revolution. In less than 10 years, this environmental awakening led to strong bi-partisan environmental legislation, such as the Clean Air Act, Clean Water Act, Endangered Species Act, Comprehensive Environmental Response, Compensation and Liability Act (Superfund), and other regulations that began to turn the tide on the environmental degradation of our one and only home.

At approximately the same time the environmental legislation was put into action, debates began to increase in scientific circles about the plausibility of a global Greenhouse Effect that may be occurring after CO_2 concentration measurements were found to be increasing in the atmosphere (resulting in Global Warming). Interestingly this debate was nothing new. Since about the turn of the 20th Century some scientists theorized that ever increasing emissions from the Industrial Revolution might someday bring Global Warming due to the 'greenhouse-like effect' that CO_2 and other gasses emitted from manufacturing plants, power plants, and other sources would cause.

In 1938, CO_2 was already found to be climbing, and the debates began. The early proponents were dismissed, and the general scientific community was skeptical, thinking this was simply an impossibility. Research continued through the 20th century that CO_2 could cause Global Warming, but much of the evidence was circumstantial. That changed in 1985, when an ice core from Antarctica that was drilled two kilometers deep captured a 150,000-year record of CO_2 in the atmosphere, as well as a complete ice age cycle of warmth, cold, warmth. Research based on that data confirmed that atmospheric CO_2 had gone up, down, up, nearly in tandem with the global changes in temperature. Scientific consensus that the Greenhouse Effect was occurring began to grow, and I myself first learned of the Greenhouse Effect and Global Warming in 1987 as I began my journey in college obtaining my first couple of degrees.

In the 30+ years since that time, further evidence was gathered, and the scientific community became in near total agreement that not only was the Greenhouse Effect occurring, but the resultant Global Warming was occurring (all you have to do is watch the news for more and more high temperature records being broken both on ground and in the sea). However, those outside of the scientific community continued to be skeptical and bickered, and nothing was done to stem the release of CO_2.

Also occurring in the 1980's was a technological advance in hydraulic fracturing of tight shale formations (fracking), which later became the fracking boom that we all know of from the news. To this day there are still no sound federal regulations covering fracking, which lose a lot of methane in the process (a much more potent greenhouse gas than CO_2). Still, the bickering and half-measures regarding stemming global warming continue, to the point where the amount of CO_2 in the atmosphere is the highest it has been in over 3 million years! We now have visually observable evidence of the ultimate result of global warming – Climate Change.

The Climate Change we are currently experiencing is just the beginning, as the latent effect of this increased CO_2 in the atmosphere means that Climate Change is going to continue to increase in strength and volatility. As ecosystems are affected, sensitive species will become endangered or extinct. Other species will have no remaining ecosystem which could result in a large-scale extinction event and species die-off that mankind has never seen. Climate change will also displace a multitude of humans on a global scale. It will cause economic instability and may lead to resource wars (as it affects our ever-depleting resources and ability to grow enough food to feed an ever-expanding population). Yes, there will be strife for many, now and for future generations. There are many issues that we as humans disagree about. But climate change is the great equalizer. It does not care if you are rich or poor, what religion you practice, or what political affiliation you associate. It will affect us all.

But hope is not lost in all of this gloom and doom, not by a longshot. The solutions presented in this handbook provide the pathway to recovery, and are simple and straightforward. If we all just take care of the patch of dirt around us, make sure to stay informed from reliable news sources, actively vote for individuals who take environmental laws and climate change seriously, and each of us support our neighbors and local communities in doing the same, we can keep climate change from becoming much worse (and possibly avoid global disaster). Our recovery to a healthy stable global climate will also return sooner, and the rate of recovery will increase as more and more Earthlings get on board.

We may only have a minor effect as individuals to slow or stop the progression of climate change and everything that comes with it, but collectively we are mighty!

Steven Sutherland, M.S., R.G., P.G., C.E.M., EE Science Officer

Made in the USA
Lexington, KY
14 December 2019

But hope is not lost in all of this gloom and doom, not by a longshot. The solutions presented in this handbook provide the pathway to recovery, and are simple and straightforward. If we all just take care of the patch of dirt around us, make sure to stay informed from reliable news sources, actively vote for individuals who take environmental laws and climate change seriously, and each of us support our neighbors and local communities in doing the same, we can keep climate change from becoming much worse (and possibly avoid global disaster). Our recovery to a healthy stable global climate will also return sooner, and the rate of recovery will increase as more and more Earthlings get on board.

We may only have a minor effect as individuals to slow or stop the progression of climate change and everything that comes with it, but collectively we are mighty!

Steven Sutherland, M.S., R.G., P.G., C.E.M., EE Science Officer

Made in the USA
Lexington, KY
14 December 2019